SUCKLED CALF PRODUCTION

by

Richard Fuller

Manager, J S R Farms Ltd, Givendale,
Pocklington, East Yorkshire

CHALCOMBE PUBLICATIONS

First published in Great Britain by
Chalcombe Publications,
13 Highwoods Drive, Marlow Bottom, Marlow, Bucks SL7 3PU

July 1988

ISBN 0 948617 13 6

Printed in Great Britain by Cambrian News Ltd, Aberystwyth

CONTENTS

Page

FOREWORD

Richard Fuller is an exceptional suckler herd manager at JSR Farms on the Yorkshire Wolds. He is a talented stockman of course, but in addition he has recognised the importance of records in pinpointing strengths and weaknesses of the enterprise so that progressive improvements can be made to herd performance and profitability.

Fortunately for the readers of this book he has a missionary's urge to write down his experiences so that others can share them. Moreover, through his excellent photography he enables readers to 'visit' the farm whilst studying his account of successful suckler herd management.

In any suckler herd the key to financial success is high calf sales per cow, made up of a high percentage of calves reared, heavy weaning weights and the leaness and shape which produce top quality carcases at the end of the day. Richard Fuller presents a detailed analysis of the components of high calf sales per cow including the importance of a compact calving period, feeding cows to achieve target body condition scores throughout the year and the influence on performance of the quality of breeding stock, especially herd sires. Having achieved high output per cow, the next step is to combine this with high stocking rates through controlled grazing management and the careful integration of grazing and conservation.

The publication of this excellent practical book could not have been timed better, coinciding as it does with a long overdue suckler herd expansion in Great Britain.

David M. Allen
Head of Beef Improvement Services
Meat and Livestock Commission

PREFACE

In 1974 I arrived at Givendale Farm, on the edge of the Yorkshire Wolds, to develop a suckled calf enterprise based on 131 hectares of unploughable grassland on the farm.

I recognised that most lowland farmers viewed suckled calf production as a 'Cinderella' enterprise. Poor returns were often the consequence of insufficient management control of the suckler cow. However, I was confident that, with effective supervision, a suckler herd would make a useful contribution to farm income.

The management techniques which I have developed at Givendale have been rewarded by significant, if not dramatic improvements in herd performance. For example, calf weaning weights increased by an average of 30 percent between 1978 and 1987. Since Givendale was awarded a Grass to Meat Award in 1983 by the Meat and Livestock Commission (MLC) and the British Grassland Society (BGS), a great deal of interest has been shown in the suckler herds. I have hosted many farm visits showing beef producers from all parts of the country how I tackle the task of making money out of suckled calf production. The keen interest which breeders have shown in my ideas and methods has prompted me to write this book.

Profitable suckled calf production requires firm positive answers to two important questions:

* Can the capital outlay required to set up a suckler herd be serviced and justified?

* Is there sufficient commitment and ability to make the project successful?

The different suckler systems which have been developed by producers have different management implications, particularly with regard to capital and labour requirements. Choosing the right suckler system to fit the farm resources is vital.

This book is based on my practical experience gained managing three suckler herds over 14 years. I hope that it will provide information and ideas so that the correct choice of suckler system is made and the herd is then managed in the most effective way.

Richard Fuller
June 1988

CHAPTER 1

PRINCIPLES OF PRODUCTION

Making a profit from suckler cows hinges on the ability to organise each phase of the production cycle into an integrated, efficient system. But before looking at each phase in detail, it is important to have a clear understanding of the principles of suckled calf production, which are essentially the same no matter which system is chosen and no matter which resources are available on a particular farm. The four basic principles are:

(i) It is essential to rear as many calves as possible in relation to the number of cows put to the bull.

(ii) The calves must weigh as much as possible at weaning, within the constraints of the production system adopted. They must also be good quality animals.

(iii) It is important to use grassland efficiently in relation to stocking rates and fertiliser usage.

(iv) The economic use of feed inputs, especially concentrates, is essential.

The calves produced from suckler cows must be regarded as a 'crop' and, as with any other farm crop, output and returns should be maximised. A high level of performance from the herd is the result of both appropriate breeding policies and effective management strategies.

BREEDING POLICIES

The choice of the most efficient breeds of cattle with the highest production potential has a significant effect on the quality and the weight of the calf crop. Performance of the suckler herd can also be considerably enhanced by a crossbreeding programme.

Figure 1.1 illustrates the advantages of using cross-bred suckler cows and suggests the characteristics to select when choosing the terminal sire. Aberdeen × Friesian cows mated with good quality Charolais bulls has proved to be an excellent combination of breeds to produce a profitable calf crop at Givendale. Other breeds, however, also perform well and keen interest is being shown at present in dairy-bred cows sired by Limousin and Simmental bulls.

FIGURE 1.1 Characteristics of suckler cows and bulls which contribute towards maximising the calf crop at Givendale

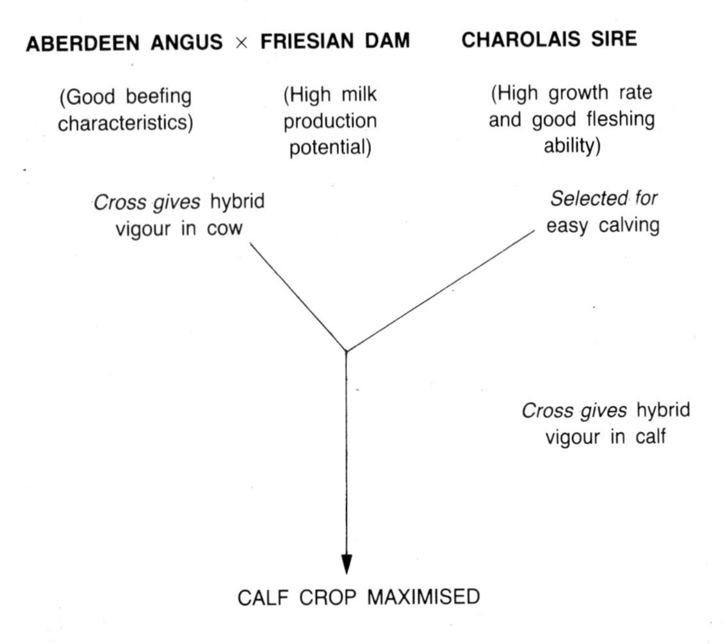

Hybrid vigour

Crossbred cows perform more efficiently than do pure bred cows. It is well known that the hybrid vigour generated by crossbreeding achieves significant advantages in terms of improved fertility, ease of calving, durability of the dam, and increased milk production which leads to better calf performance. Certain breeds combine together better than others, but if crossbreeding is repeated after the first cross then the advantages of hybrid vigour decline. It is important to recognise that the effect of hybrid vigour is enhanced by breeding crossbred cows to a third breed of sire.

Farmers are currently involved in fairly complicated crossbreeding programmes with the aim of reducing the influence of Holstein blood in their suckler herds. Time will tell how successful these programmes are, but the basic principles of crossbreeding remain the same. Care must be taken not to breed less efficient cows, and greater attention paid to the conformation of the sire used on suckler cows bred from Friesian/Holstein cows may well prove to be the most efficient way of improving the shape of the calves.

Limousin bull with Angus × Friesian heifers.

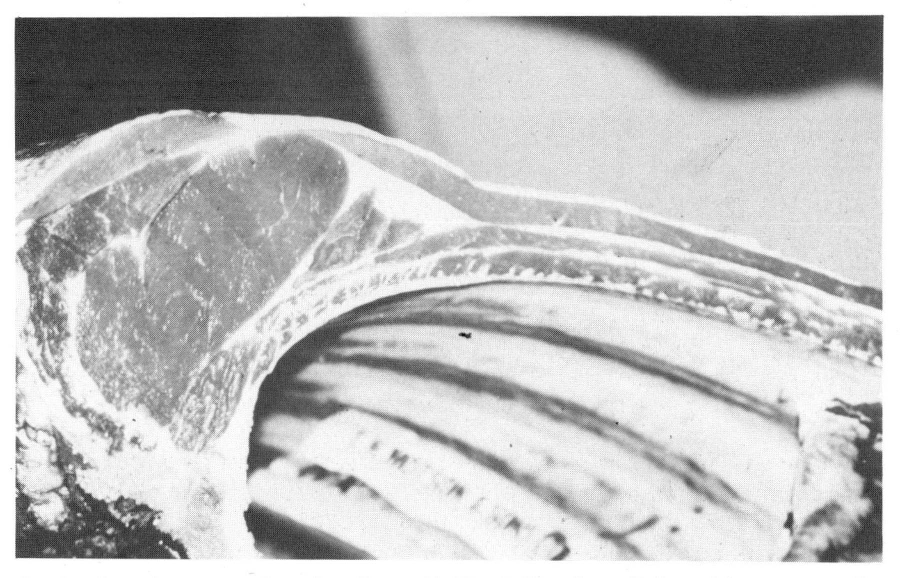

Section through carcase of good quality suckled beef. Note the well-shaped, lean eye muscle showing little intra-muscular fat, and the lean meat present on the rib. A good quality carcase is strongly influenced by the choice of sire.

Choice of breeds

The choice of suckler cow breeds depends on the resources and topography of individual farms. Dairy bred cows are usually used in lowland and most upland situations because of the large numbers available from which to select good breeding types. On the other hand, the harsh environment which is associated with hill areas demands the use of specific, more hardy breeds of cows such as the Galloway and the Welsh Black.

The relatively small Hereford × or Angus × Friesian cows have proved to be very efficient producers of beef calves on many farms, especially where resources are scarce. Breeding efficiency and milking ability are good, maintenance requirements are modest and, providing the calves are sired by the right type of bull, production from these cows is excellent in terms of calf size and quality.

Larger cows, bred by using heavy continental bulls, require more maintenance and therefore cost more to keep. Stocking rates are reduced both in yards and at grass, and because their value is higher than that of smaller cows capital investment increases, although a high cull cow value is advantageous. Calves produced by larger cows do not necessarily increase in weight in proportion to the increase in cow weight, therefore the value of the calf crop from larger cows is unlikely to justify the extra costs involved in production. While there is probably a case for keeping larger cows where resources allow them to express their full potential, they are not suitable for the majority of upland situations.

MANAGEMENT STRATEGIES

Two key features of profitable suckled calf production are that every cow should produce a calf each year, and the growth rates of the calves should be maximised. Both of these targets must be borne in mind when considering which management practices to adopt.

Figure 1.2 shows management practices which play a part in the process of maximising the calf crop from the Givendale herds.

FIGURE 1.2 Management practices to maximise the calf crop

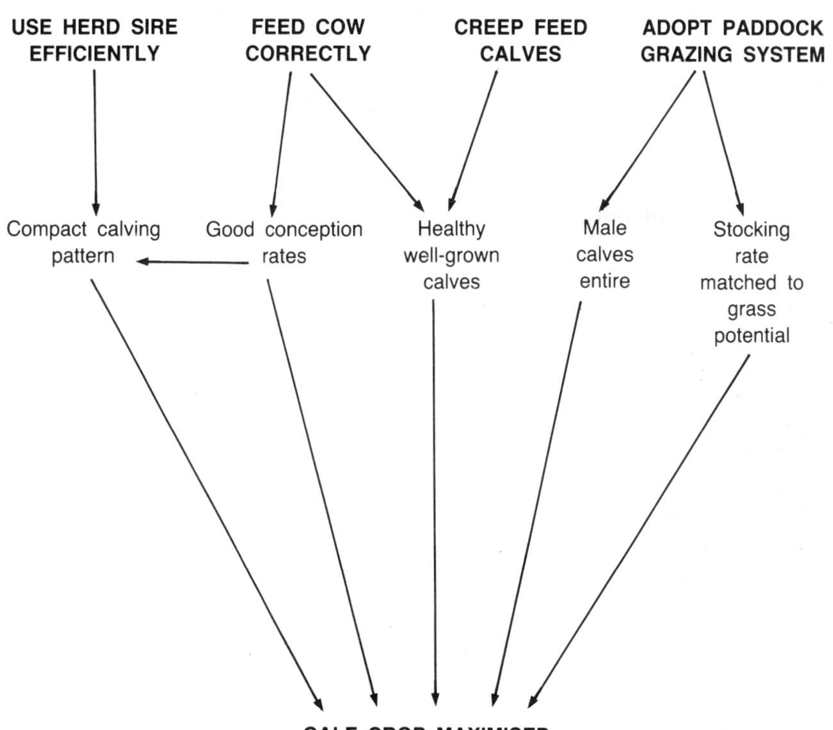

Understanding the nutritional requirements of the cow at different stages of the production cycle and feeding her correctly at each stage ensures that she both produces a healthy calf and is in-calf again within 3 months. Growth rates of calves in the suckler herd are maximised firstly by ensuring that the cows are fed as required to produce adequate amounts of milk, and secondly by offering the calves creep feed. At Givendale, the basis of the efficient management of both cows and calves to maximise calf output is the achievement of a compact calving pattern.

Profitability is also increased by paying close attention to the cost of feeding the herd and identifying areas where savings can be made without adversely affecting cow or calf performance. Finally, the adoption of a paddock grazing system not only allows the efficient use of grass, but also enables male calves to be kept entire, thus increasing their potential growth rate.

Good quality suckler cow with her 1 month old calf, born in September, at foot.

The same cow and calf 9 months later. The cow (485 kg) is due to calve again within the next 2 months; her calf now weighs 480 kg.

Thriving February-born calves at grass in early June. A compact pattern of calving allows the production of even groups of calves, and easy management of cows and calves.

COMPACT CALVING PATTERN

As already noted, a compact calving pattern has a dominant role to play in the success of a suckler herd. A calving spread of not more than 9 weeks should be the aim, although an even shorter period would be advantageous. It is crucial, however, to get the suckler cow back in calf within 12 weeks of calving, as failure results in disruption to herd management which in turn leads to lost production.

Figure 1.3 illustrates the main advantages of a compact calving pattern, all of which contribute to maximising herd output.

FIGURE 1.3 Advantages of a compact calving pattern

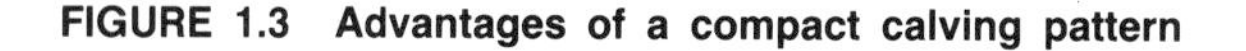

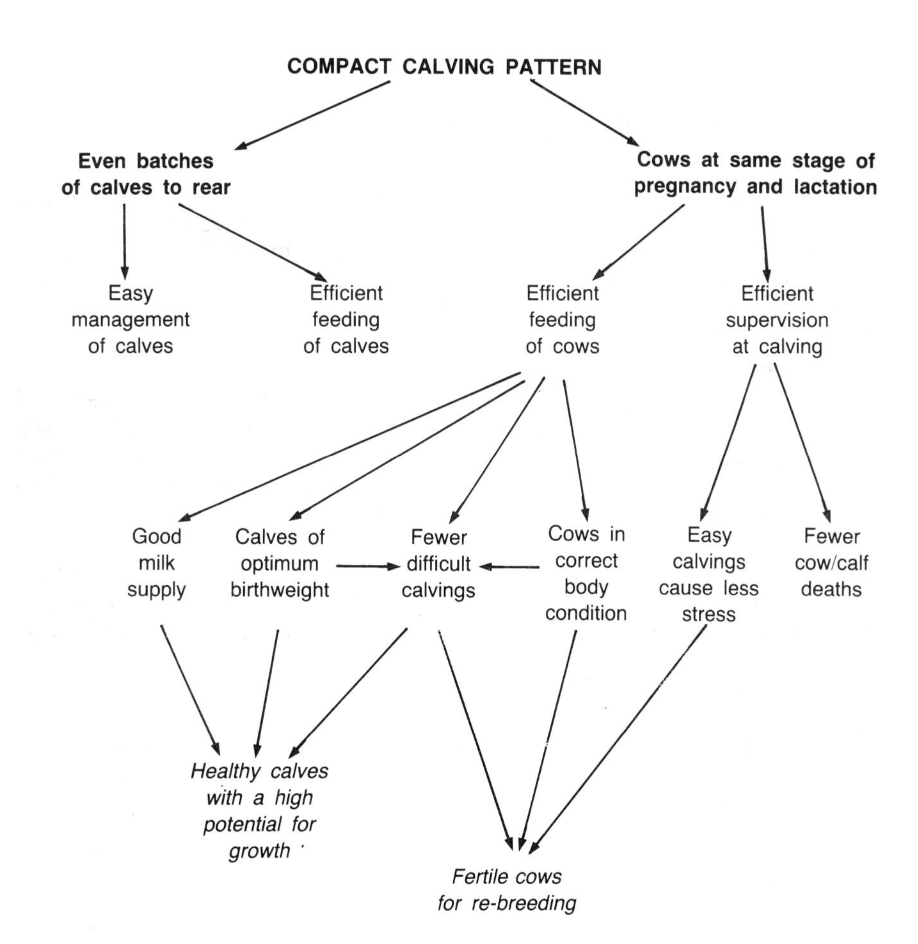

The key to achieving a compact pattern of calving is knowing the right level of feeding and the correct body condition for the cow at the critical stages of the reproductive cycle, as discussed in Chapter 3. A second important factor which also influences the calving pattern is the efficient use and performance of the herd sire, and this aspect of suckler herd management is considered in more detail in Chapter 5.

CHAPTER 2

SYSTEMS OF PRODUCTION

Systems of suckled calf production are usually first identified by the time of year when the cows calve. The most common systems involve suckler cows calving at grass in the autumn, in buildings during the winter, or at grass in late spring (May and June). In a few cases summer calving (i.e. calving in July and August) is practical, but production tends to be poor in terms of output per cow.

The season of calving has a number of implications for the organisation of the herd. Therefore, when planning a suckler system the first question which must be asked is "When would be the best time to calve the cows?" Serious thought must be given to this question, and to arrive at the correct answer the resources of each individual farm must be examined.

The first part of this chapter highlights some important factors which influence the choice of calving season. In the second part, the organisational implications of autumn, winter and spring calving systems are considered.

CHOOSING THE CALVING SEASON

The main criteria that influence the decision when to calve are:

* Feeding
* Housing
* Labour
* Calf disposal

Feeding

As we shall see later, the majority of management decisions concerning the suckler herd are taken with feeding in mind because the costs associated with feeding have a major impact on performance, and therefore profitability.

Silage is probably the most common form of forage fed to suckler cows, and the relatively high costs which are associated with silage-making mean that full consideration must be given to using this feed economically. Silage may be derived from the early cutting of permanent grass which is part of the grazing area later in the season, or from specific temporary leys grown as a break crop in an arable rotation. The latter option has the disadvantage of

Outwintered Luing cattle on the Isle of Islay in January.

taking land out of the arable cropping programme, thereby increasing the cost of the silage because the ley has to carry its share of the arable overheads. Advantages, however, are that a ley produces heavy cuts of grass and it also provides a 'clean' grazing area for weaned calves in the autumn.

When considering suckler cows as part of a lowland arable system, the opportunity arises to use arable by-products such as sugar beet top silage and stock feed potatoes, or even to grow a specific high-yielding crop such as fodder beet. The most obvious arable 'waste' product, however, is straw, of which thousands of tonnes are produced each year.

It has been shown that untreated wheat straw, supplemented by a small amount of cereal, is an adequate feed to over-winter May or June calving cows. If the wheat straw is treated with ammonia the resulting increase in digestibility ("D" value) also enables lactating cows to be fed successfully on a straw-based diet, with cereal supplementation. Straw, either treated or untreated, therefore has a major role in suckler cow production as a relatively cheap form of forage in most situations, although transport costs generally make it a more expensive commodity on hill farms.

Housing

Winter housing of suckler cows is desirable in the majority of cases because damage to land is avoided, and because housed animals require considerably less energy than do out-wintered ones. There may be a case, however, for out-wintering spring calving cows on the milder west side of the country when well-drained land is available.

Suckler cows can be housed in a wide range of buildings. The availability of straw for bedding is an important factor when considering the type of housing, but it must be stressed that expensive and elaborate buildings cannot be justified in the majority of cases. The principle of housing suckler cows is to give them plenty of air and a dry sheltered lying area.

Open-fronted pole barn sheds are ideal, but if straw is scarce and expensive then cubicles or cow kennels are suitable provided the cows do not have to calve in them. Slurry disposal may be a problem with these latter types of housing which use little straw.

When planning winter accommodation it is important to remember that ideally groups should consist of no more than about 40 cows in one yard. There are several reasons for this:

(i) If the cows are to be mated during the winter period then 40 cows are enough for one bull to cope with.

(ii) Calves, when housed with their dams, tend to perform better in smaller groups and creep feeding is easier to regulate.

(iii) With larger groups of cows and calves there will inevitably be a percentage of animals that succumb to environmental stress, which will result in poor performance.

It is also important to be able to separate the breeding herd into groups depending on the nutritional needs of the cows. Ideally the herd should be split into three groups:

(i) Replacement heifers
(ii) Second calvers and thin cows
(iii) Mature cows.

Building design and feeding systems should allow the economic use of labour and should be sympathetic towards the well-being of the livestock. Stressful conditions, such as badly-ventilated buildings or insufficient feeding space, will contribute towards depressed performance. Buildings which create labour-intensive situations may lead to jobs not being done properly, to lack of attention to detail and to inadequate observation in a rushed situation.

Cubicle house for suckler cows.▶

◀ Open fronted housing creates a healthy environment for suckler cows.

Autumn-calved cows and calves housed in a well bedded straw yard. ▶

Labour

On arable farms a danger is that the suckler herd is treated as a minor part of the business because it is there to utilise an area of rough grazing or some parkland. The herd may well not be big enough to justify a full-time stockman, and the lack of specific responsibility may lead to management practices being delayed or not carried out at all.

This has the effect of creating a downward spiral of poor performance which wastes farm resources and can result in the business actually making a loss. It is therefore important to appreciate the need to identify the peak labour requirements of the suckler herd and to plan accordingly.

Supervision during the calving period demands the most concentrated labour input of the year, therefore it is advisable to calve the cows during a relatively slack period on the farm. For instance, it would be unwise to calve during the harvest period on an arable farm unless there was labour specifically available for the suckler herd. In the upland situation it may be inadvisable to allow calving to coincide with the lambing period because performance of both operations could suffer.

It is also important to recognise the need to spend extra time with the herd during the mating period to ensure that the bull is working properly. Assuming that all is well at such a critical time is not good enough, and may lead to catastrophic losses if a bull fails to serve the cows or if his fertility drops.

Calf disposal

It is desirable to maximise the output from each cow in the herd, and in general the older the calf when it is sold, the higher its value. Autumn-born calves secure the highest prices when sold the following autumn. Calves born in the same year as they are weaned are lighter, so to maximise the return per cow the calves need to be housed during the following winter for a growing-on or finishing period. In a lowland situation, therefore, where arable by-products and adequate buildings are available, selling finished cattle rather than weaned calves should be considered.

On the other hand, in a hill situation forage may be scarce and expensive, and there is a danger that the cows compete for resources with the weaned calves. This can result in neither the cows nor the calves being fed adequately, and the sale of weaned calves before winter may then be preferred.

These examples illustrate that careful planning and budgeting is necessary to identify the best method of calf disposal to suit the resources of individual farms.

AUTUMN CALVING

Autumn calving in September, October or November is a high input/high output system. Because the cows are calving towards the end of the grazing season, they tend to be fairly fit, and possibly over-fat, so supervision during the calving period is demanding.

Once housed the herd has to be fed well, not only to cater for lactation but also to ensure that good conception rates are achieved.Stocking density in the buildings is lower than for spring calvers to allow extra space for the growing calves and for their creep area. The labour requirement for an autumn calving herd during autumn and winter is relatively high, although this situation may be advantageous if spare labour from the arable enterprise can be utilised during the winter.

At the end of the winter the cows and calves may be turned out to grass together, but stocking rates should make allowance for the area required by the fairly large calves. Alternatively the calves may be weaned and then either turned out on to good 'clean' grassland, or kept inside for a finishing period. In this latter situation the cows may be turned away on to inferior grazing and stocked tightly until they calve again.

Weaning the calves at turnout particularly suits those farms where an area of rough hill is available for the cows. Generally, however, in lowland situations the herd is best turned out in its entirety in order that, in addition to grazing good quality spring grass, the calves can benefit from the flush of milk from the cows. The high output of the system is realised when heavy calves are weaned in early August.

WINTER CALVING

Winter calving in buildings during January, February and March falls between autumn and spring calving in terms of input and output. On some farms the main drawback to the system is an increased risk of disease, particularly scouring, in young calves—but with careful management and the use of modern vaccines the rearing period should be relatively trouble-free.

Winter calving systems have the advantage of allowing the turnout of fairly strong calves which are able to take full advantage of a flush of milk from their dams. The cows can be fed more cheaply through the winter than can autumn calving animals and they should be stocked more tightly at grass after turnout because they have smaller calves. The calves, when weaned in October or November, are 3 months younger than autumn born calves, and they weigh approximately 65 kg less, but because of lower feed costs and higher stocking rates returns from the two systems are similar. After weaning the calves may be sold or, alternatively, if buildings and feed are available, retained and finished. This latter option ensures maximum output from the herd.

February calving suckler cows and 3 month old calves turned out with a stock bull.

Autumn calving herd: new born calf at grass in late September.

Winter calving herd: new born calf in straw yard.

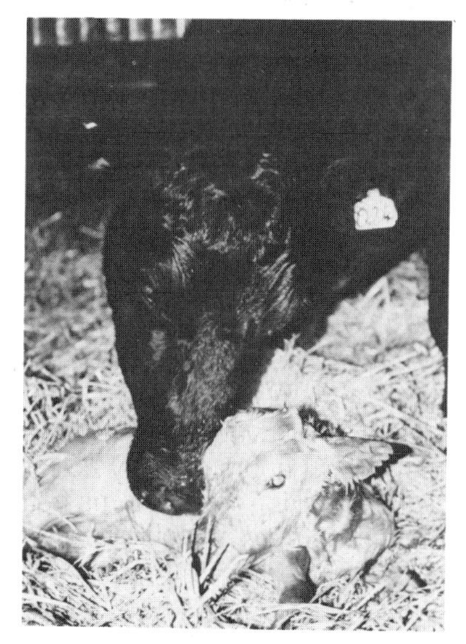

SPRING CALVING

Spring calving in April, May or June is a low input/low output system which may suit farms where large quantities of straw are available. The dry cows can be fed relatively cheaply during the winter either on ammonia-treated straw alone, or on untreated straw plus a small amount of supplement. The cows calve outside, having lost weight during the winter so calving difficulties are minimised. After calving, the availability of grass which is of a high feeding value means that the cows conceive successfully as their body condition improves.

The biggest drawback to this system is that calves are still relatively small by the time they are weaned in the autumn. To make the system economically viable, after weaning the calves must be housed either for finishing rapidly or for feeding-on as stores, which can then be sold as and when resources dictate.

A spring calving system has been developed at High Mowthorpe Experimental Husbandry Farm (EHF) specifically to demonstrate the utilisation of unploughable or marginal grassland on arable farms. The cows are wintered cheaply on straw plus a minimum level of supplementation, and the system makes use of the ability of cows to 'live off their backs' in winter and to gain condition at grass during the summer.

The High Mowthorpe strategy highlights the need to produce good quality calves which are put on to a fast finishing regime after they have been weaned in November. Table 2.1 shows the performance of Charolais-sired calves in 1985/86.

TABLE 2.1 Performance of spring born suckler calves finished at High Mowthorpe EHF on different feeding regimes during winter 1985/86

	Heifers	*Bulls*	*Steers*	*Bulls*	*Steers*
	Ad-lib silage + barley at			*Ad-lib barley*	
	3kg/day	*4kg/day*	*4kg/day*		
Daily liveweight gain (kg)	1.06	1.54	1.27	1.89	1.40
Slaughter weight (kg)	403	496	478	496	464
Carcase weight (kg)	212	272	254	208	253
Age at slaughter (months)	12.5	11.75	12.5	10.5	11.75

Source: High Mowthorpe EHF

CHAPTER 3

MANAGEMENT OF THE SUCKLER COW

The function of the suckler cow is to produce a good quality calf which, within the adopted production system, weighs as much as possible in relation to her own body size. This should be achieved as economically as possible every 365 days.

Three areas of suckler cow management are particularly important—nutrition, health and calving. In addition, condition scoring and pregnancy diagnosis are useful management tools.

NUTRITION

Adequate nutrition of suckler cows has a profound effect on reproductive performance and profitability of the herd. The majority of management decisions are related to feeding, and feed costs make up the highest proportion of the costs of suckled calf production. In addition serious problems, illustrated in Figure 3.1, arise when cows are underfed.

Strategies to feed a cow adequately and economically depend on her size, her body condition, the stage of her reproductive cycle and whether she is still growing. In order to ensure good performance and the cost-effective use of feed, it is important to have a clear understanding of the nutritional needs of the cow during the various stages of her reproductive cycle.

FIGURE 3.1 Consequences of the undernutrition of suckler cows

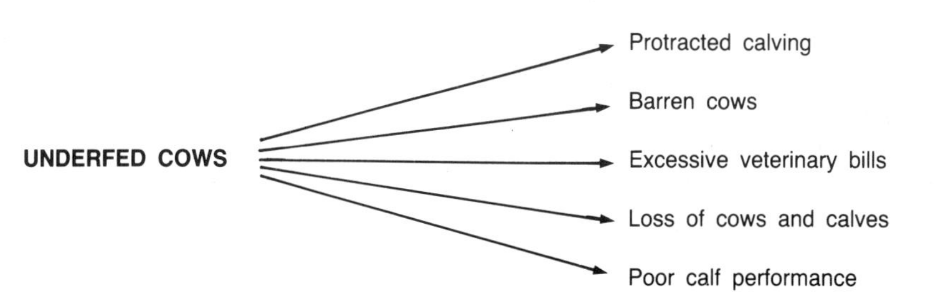

Nutrition in relation to the reproductive cycle

When we study the job that the suckler cow has to do we see that she has to maintain her own body, feed her calf, come into heat, conceive promptly and then deliver a live calf. The range of nutritional requirements for the cow to do this job well is considerable, and it is important to recognise that the nutritional resources available to her are used in a specific order. The priority of a suckling cow is to feed her calf by producing milk. Her next priority is to maintain herself and, finally, reproduction comes last on her list. So we can see that if a cow does not get adequate nourishment then reproductive performance is the first to suffer.

The most critical period in the reproductive cycle is the time between calving and mating. The energy requirement of the cow is at its highest then because she is trying to put on weight after calving, the calf's demand for milk is at its peak and she is trying to get her reproductive system ready for re-breeding. Adequate intakes of relatively high quality feeds are needed at this time, and it is also important to continue to keep energy intake at a high level for 6 to 8 weeks after mating to ensure the successful implantation of the foetus.

In mid-gestation energy requirements are much lower and cows can afford to lose body weight with no detriment to themselves, their calf at foot nor their foetus. During this stage it is important to be aware that savings in feed costs may be possible.

In late gestation cows must not be allowed to get too thin because 80 percent of foetal growth takes place during the last 8 weeks of pregnancy and, in addition, the cow is preparing herself for lactation. Breeders who worry about difficult calvings tend to run too much condition off their cows, which is a bad policy because it is difficult and expensive to feed thin cows for lactation as well as for the necessary improvement in condition which is required for successful re-breeding. Rather than regain body condition, cows may well channel the extra feed provided after calving into excess milk production, which can result in scouring calves, which in turn lead to management problems.

By contrast, the opposite situation is that cows should not be *excessively* fat at calving because calving difficulties are more likely due both to fat deposits in the pelvis and also to an increase in the weight of the calf. Emphasis, however, is placed on the term 'excessive', especially where heavy continental sires are used. At Givendale, autumn calving cows which calve at the end of the grazing season produce calves which are, on average, 7 kg heavier than calves born to February calving cows which are fed a controlled diet. However, the rate of assisted calvings in both herds is exactly the same.

Autumn calving cows eating good quality silage.

Feeding for high conception rates

Mating at grass

Cows that are mated while they are grazing are ideally suited to achieve high conception rates, provided that an adequate allowance of good quality grass is available and intakes are not restricted for any reason. In addition, cows should not have been allowed to get too thin prior to calving.

Mating in winter

The level of feeding for cows that are mated in the winter is of critical importance. To achieve the best conception results the herd must be housed in 3 separate groups, which require different levels of feeding:

(i) *First calved heifers* require extra feed in a non-competitive environment to allow continued growth as well as lactation.

(ii) *Second calvers and thin cows* also need extra feed. Second calvers are notorious for refusing to gain weight after calving. As thin cows are almost invariably the older ones, the heavy milkers or suckling twins, the need for extra feed is obvious.

(iii) *Mature cows in good condition* are better housed away from the other two groups to prevent bullying and general environmental stress. If it is not practical to house the herd in 3 groups then the first 2 groups should be housed together, away from the mature cows. Savings in feed costs can be made with the mature cow group.

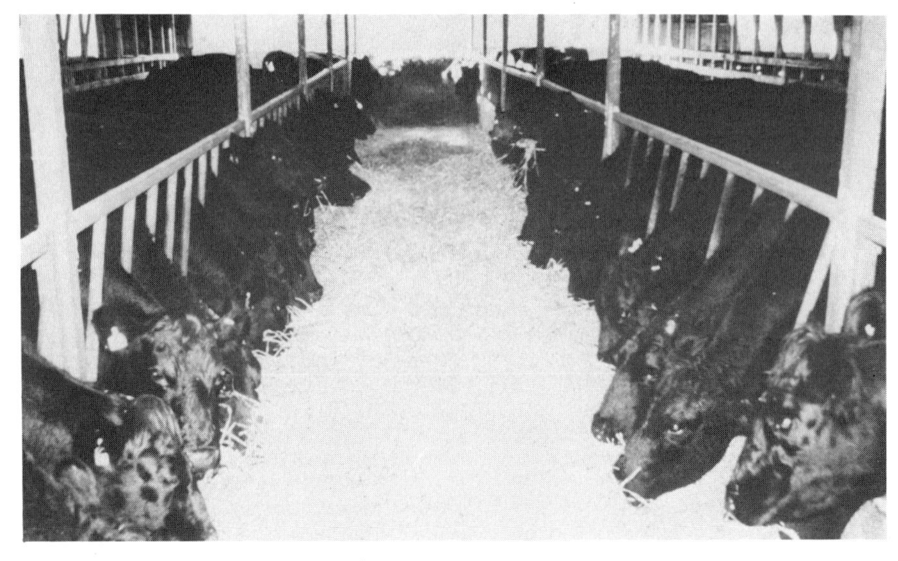

Winter calving cows eating anhydrous ammonia treated wheat straw in a cubicle shed.

It is important to recognise the need to feed cows on a rising plane of nutrition up to the mating period to allow for the increasing milk demand of the calves. It is also important to feed high phosphorus minerals to the cows because phosphorus has been shown to aid fertility. Finally it is not sound practice to change dramatically the ration of cows during the mating period because sudden changes in diet can lead to reduced conception rates.

CONDITION SCORING

Condition scoring should be used to assess the nutritional needs of cows at the critical stages of their reproductive cycle. Initially scoring should be done by handling the cows individually, but when sufficient experience has been gained it should be possible to score them by eye.

When handling the cows the procedure is to grip the loin halfway between the last rib and the hip bone. The hand is placed on top of the loin with the fingers pointing towards the spine, and the thumb is pressed round the transverse processes of the spine to feel the fat cover over the tips of the processes, as shown in Figure 3.2.

FIGURE 3.2 Condition scoring

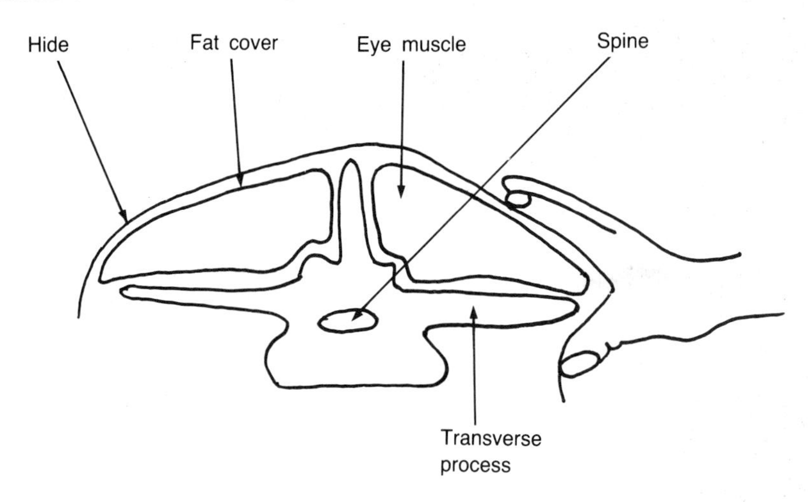

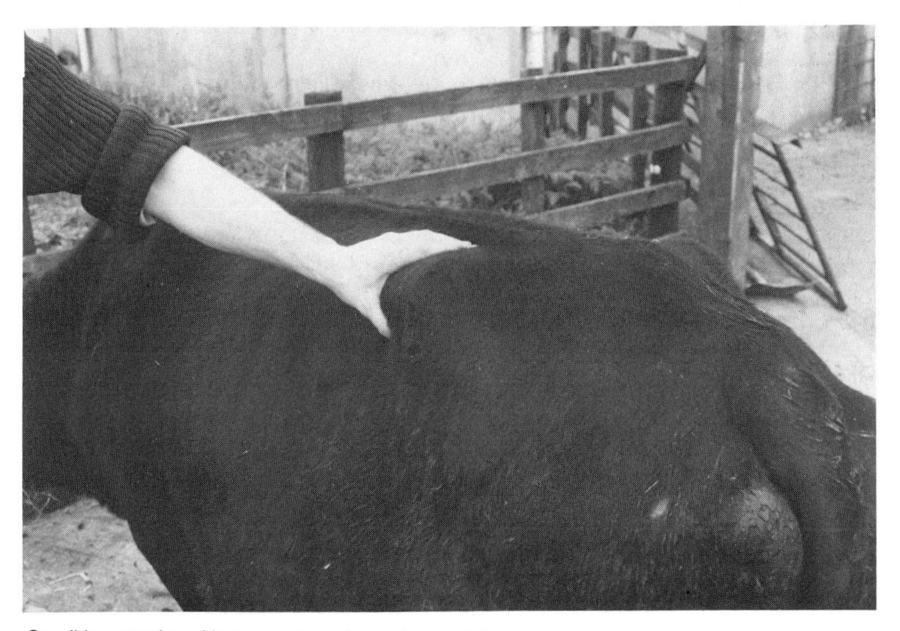

Condition scoring. Note position of hand over loin area.

Body condition is scored on a six point scale from 0 (very thin) to 5 (very fat) as follows:

Condition score	*Description of loin*
0	The spine stands out and the ends of the processes feel very sharp
1	The spine is still prominent but the transverse processes feel less sharp
2	The spine is less prominent and the transverse processes feel cushioned by the increased fat cover
3	The transverse processes can only just be felt
4	The transverse processes cannot be felt
5	The transverse processes cannot be felt under a thick layer of fat

Target condition scores

Table 3.1 shows the target condition scores for autumn and spring calving cows at different stages of the reproductive cycle.

TABLE 3.1 Target condition scores

	Target condition score	
Stage of reproductive cycle	*Autumn calving cows*	*Spring calving cows*
At calving	3	2.5
At mating	2.5	2
Mid-pregnancy	2	3

The dangers of allowing cows to become too thin or too fat prior to calving has already been noted, and identifying cow condition with a scale of figures allows a better definition of the correct body condition.

Autumn calving cows

Calving cows in the autumn at a condition score of 3 reduces the need for expensive feeds after calving and allows half a score drop in condition up to the bulling period. If the cows were to be calved at a condition score of 2, large amounts of expensive feed would have to be given to provide sufficient nutrients for lactation in addition to allowing an increase in body weight. After mating and successful implantation of the embryo, cow condition can be allowed to drop to a score of 2 in mid pregnancy, enabling savings to be made in feed.

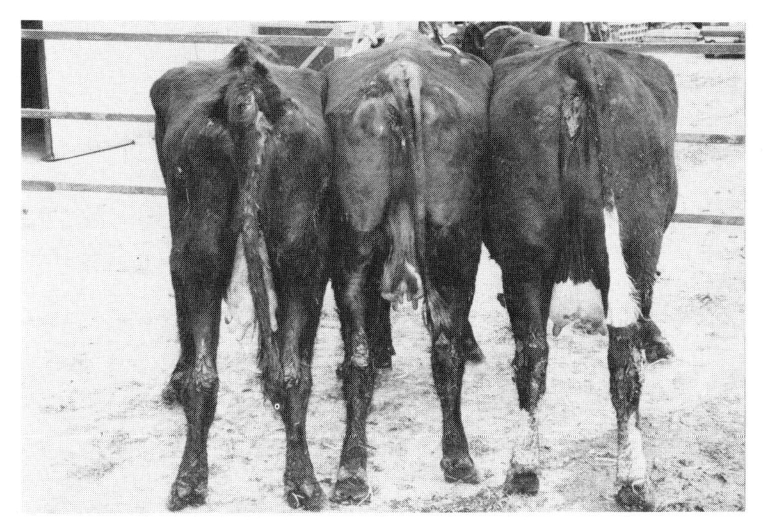

Condition score: rear view of cows in, from left to right, score 0, score 2 and score 4.

After turn out, which occurs during mid-pregnancy, body condition should increase to score 3 during the grazing season. A higher score than 3 could lead to calving problems, especially in herds where heavy continental sires are used. Very tight stocking of the cows after weaning may be necessary to stop the cows putting on too much condition prior to calving, especially in a good year for grass growth.

Spring calving cows

As we have seen, the advantages of spring calving over autumn calving are lower winter feed requirements and the utilisation of nutritious grass to achieve good conception rates. The target score for calving is 2.5 and, because a high intake of nutrients from spring grass results in a rising plane of nutrition, cows should conceive at condition score 2. During the grazing season the condition of cows should gradually improve to score 3, which allows subsequent savings in feed due to the use of body reserves through the winter, prior to calving at score 2.5.

Advantages of condition scoring

Condition scoring helps to:

* Maximise conception rates
* Allow economic use of feeds
* Identify thin cows which can then be given preferential treatment
* Avoid calving problems.

HERD HEALTH

Certain diseases and conditions are of particular importance where suckler cows are concerned and it is worthwhile discussing at this stage the most common problems. Suckler cows do not enjoy the individual attention that dairy cows receive twice a day as they go through the milking parlour, which means that it is important to be constantly observant for diseases such as mastitis. An alert stockman can do much to minimise the effects of disease.

Stress is a prime catalyst for triggering many conditions, so stressful situations such as rough handling or badly ventilated buildings should be avoided. When a disease can be controlled by adopting preventive measures, they should be used because prevention is more economical and humane in the long term than treatment or the loss of animals. It must also be remembered that drugs should not be relied upon to compensate for bad management practices—they should only be regarded as an aid to sound stockmanship.

Common diseases and disorders

Hypomagnesaemia (staggers)

This condition, which is due to inadequate magnesium in the blood stream of the cow, is a killer. Lactating cows or cows whose calves have just been weaned are at the greatest risk. Hypomagnesaemia occurs most frequently when cows are turned out onto lush spring grass, but it may also occur when cows graze a flush of grass in the autumn. Stress associated with weaning or moving cows, sudden drops in temperature and poor weather conditions all seem to be catalysts for the development of this disorder.

Prevention is by far the best course of action to take because the first symptom is often a dead cow. The feeding of magnesium-rich supplements during the high risk periods helps to alleviate the problem. When a cow suffering from hypomagnesaemia is found alive she must be treated very quickly with an intravenous injection of magnesium salts.

Milk fever (hypocalcaemia)

Milk fever, which occurs in cows before, during or more commonly after calving, is caused by a fall in calcium, and possibly phosphorus levels in the bloodstream. It is not possible to take preventative measures so swift action is necessary to treat affected animals. This takes the form of an intravenous or sub-cutaneous injection of a mixture of calcium and phosphorus.

Cow taking magnesium syrup, supplied at turnout to grass to prevent hypomagnesaemia.

Older cows which milk well are most susceptible, and animals which have had an attack are liable to a repetition when they next calve. The incidence of the disease appears to increase when cows calve outside in cold, wet weather, and in these cases hypomagnesaemia may also be involved to some extent.

Mastitis

This is a very nasty condition and in the worst cases cows can lose their milking ability, abort or even die. The treatment of mastitis can be extremely time consuming and is not always successful. Quarters can still be lost after several weeks of stripping out the udder and giving antibiotic therapy.

The symptoms to be on the look out for in suckler cows are:

* Clots hanging from the end of a teat.
* Swollen quarters—special note should be taken of cracked teats which are too sore for the cow to allow her calf to suckle from, as this situation often leads to mastitis.
* A cow which kicks at her calf when it tries to suckle may have a painful, infected udder.
* Lameness in the back leg may be caused by the pain of an infected udder.
* A cow which stands away from the rest of the herd and is obviously ill must be suspected of having mastitis.

There are many different types of bacteria which cause mastitis; some are more virulent than others. Flies may be responsible for spreading certain types of infection, such as summer mastitis, from cow to cow.

The following preventative measures can be taken, although they are not foolproof:

* Drying cows off after weaning is a critical time in the control of mastitis. The udder should be treated with a long acting 'dry cow' antibiotic and the teats should be washed and dipped with a proprietary teat dip. The cows should also be sprayed with an anti-fly preparation, which can be done using a knapsack sprayer.
* Housing the cows on straw after weaning for 3 to 4 days helps to stop milk production, so reducing the risk of an attack of mastitis.
* Anti-fly ear tags or pour-on anti-fly preparations are very useful to keep flies at bay where they are a problem, such as in wooded areas or near water. These products are particularly useful in protecting maiden heifers against summer mastitis.

Infertility

The inability to breed regularly causes infinitely greater financial losses than those for all other diseases put together. The immediate loss is not so apparent as it is with a specific disease, but delay or failure to breed interupts production and therefore reduces profit. There are many causes of infertility, but the most common ones are under the control of the stockman. They include:

* *Poor feeding and body condition.* As explained earlier, cows must be in good condition at mating. They must also be given an adequate supply of minerals, which must be high in phosphorus especially during the mating period. Copper also has an effect on fertility and deficiencies should be rectified on the advice of a veterinary surgeon.

* *Infection of the uterus.* This can occur after a cow has retained the afterbirth or when attention to cleanliness at calving has been inadequate.

* *Infection of the ovaries.* This generally manifests itself as irregular or absent heat periods.

Mating records are very useful to identify cows which have a breeding problem. If cows do not hold to 2 services or if they are not seen bulling at all then they should be examined by a veterinary surgeon. Appropriate action can then be taken.

Administration of a copper bolus to correct copper deficiency.

Parasites

Parasitic infections interfere with the performance of suckler cows and should therefore be controlled. The efficiency of modern drugs makes control simple and economic.

Worms should be treated by using a broad spectrum anthelmintic on suckler cows shortly after housing in the autumn. This treatment rids cows of all types of worms and also kills hibernating larvae, so preventing pasture contamination the following spring. Due to the fact that modern anthelmintics are so efficient, once-a-year treatment should be adequate to prevent a build-up of all types of worms in the herd. Cows or heifers which are "bought-in" should always be wormed as soon as they come onto the farm to avoid contamination of pasture and a breakdown in the system.

Lice are common in untreated cattle during the winter months. There are several preparations on the market which control this parasite and they should be applied soon after cattle are housed in the autumn. However, infestation may re-occur by late February, in which case the cows should be dusted with louse powder to eliminate itching and hair loss.

PREGNANCY DIAGNOSIS

The confirmation that, following mating, cows are pregnant is useful for several reasons. In particular it is an early warning guide for:

* Planning herd replacements
* Efficient cull cow marketing
* Identifying a sub-fertile bull
* Identifying fertility problems in cows.

In spring calving herds cows should be pregnancy tested shortly after weaning in the autumn. Barren cows can then be identified and either sold immediately or fattened for sale at a later date. Autumn calving cows are best examined just before turn out.

When spring and autumn herds are run on the same farm, non-pregnant cows can be moved from one herd to the other. Pregnancy testing is necessary in this situation to ensure that non-pregnant cows can be re-bulled at the correct time. Only younger cows should be considered for this procedure, and then only after a satisfactory internal examination by a veterinary surgeon. Older cows and those with recurrent fertility problems should be culled.

CALVING

The profitability of the suckler herd depends on the cows giving birth to live calves and every effort should be made to ensure that this happens. The target must be for every cow to have a live calf. The majority of cows calve quite normally without assistance, but it is never possible to predict which cow is likely to need assistance, so constant observation is necessary during the calving period.

If calving occurs outside, every effort should be made to make the stockman's job as easy as possible. The calving field should be as close to his home as possible and it should be accessible with a vehicle. Night-time patrols are more conveniently carried out with a vehicle, which should be equipped with good lights. If cows have to be calved a long way from the farm buildings then it is well worth constructing a catching pen in the field.

Once calving is in full swing the herd should be checked every 3 to 4 hours around the clock. The majority of calves are lost between midnight and 6am because the 3am check is the one which is most often missed.

Emphasis should be put on achieving and maintaining a tight calving period, with a target of 85 percent of cows calving in the first 4 weeks. Quite clearly this situation helps to make monitoring the herd easier.

Wooden catching pen for use when assistance with calving is required. This pen is placed with access from 3 fields.

Cow showing water bag prior to calving.

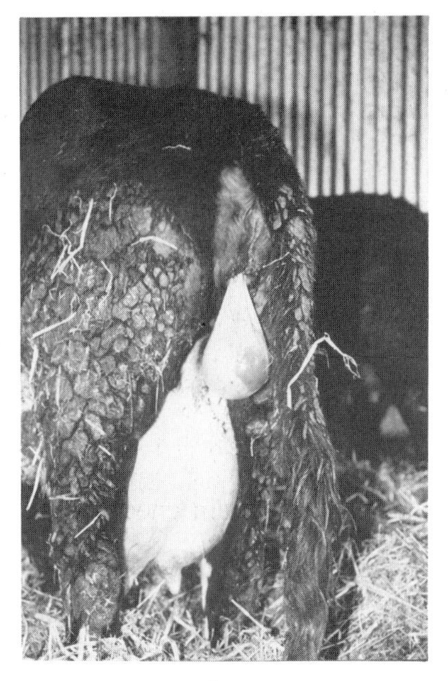

The breed of bull used on the herd has an effect on calving problems, as shown in Table 3.2, but problems can arise with all breeds. If the herd is managed properly, there is no reason why calf mortality should be any higher in a herd where a Charolais bull is used than in one where an Angus bull is the chosen sire. There is certainly more skill and attention to detail required with a Charolais bull but this is well rewarded by the extra value of the calf crop.

TABLE 3.2 Assisted calvings and calf mortality for different breeds of sire

	Assisted calvings (%)	*Calf mortality (%)*
Sire breed		
Charolais	9.0	4.8
Simmental	8.9	4.2
South Devon	8.7	4.0
Devon	6.4	2.6
Limousin	7.4	3.8
Lincoln Red	6.7	2.0
Sussex	4.5	1.5
Hereford	4.0	1.6
Angus	2.4	1.3

Source: Meat and Livestock Commission

Stages in calving

A thorough understanding of the physical act of calving is very helpful in achieving good results in practice. The stages in calvingcan be split theoretically into four, but it must be recognised that calving is basically a continuous process.

1. *Preliminary stage*

This usually starts several days before calving takes place. The vulva becomes swollen and the udder increases in size and becomes tender. The ligaments on each side of the tail head slowly slacken until calving is imminent, when they can no longer be felt.

2. *Dilation of cervix*

When this stage starts the cow becomes uneasy and usually separates herself from the rest of the herd. As dilation progresses the cow assumes the typical stance of arched back and tail held straight out. Milk often drips from her teats and she repeatedly lies down and gets up. The water bag then appears. She becomes more restless, and when the water bag bursts she licks the fluid off the ground and often calls to the calf.

The use of a calving aid for a normally presented calf.

Calving: the use of calving ropes for a normal presentation.

3. *Expulsion of the calf*

When the cervix is fully dilated the contractions become more intense and more frequent. Each contraction pushes the calf a little further up the pelvic canal until the front feet and then the head are pushed out, followed by the rest of the calf. The sooner the cow starts to lick her calf the better because the licking process stimulates and dries the calf.

4. *Expulsion of the afterbirth*

This usually happens within 1 or 2 hours of calving. Sometimes the afterbirth is retained, however, and it has to be pulled away after about 4 days. As this can lead to infection of the uterus, veterinary advice should be sought.

Calving problems

Calving is not always straightforward, and knowing when and how to assist is critical in saving the lives of cows and calves. The first step that all stockmen should take is to acquire a calving aid and then learn to use it correctly. This simple, inexpensive piece of equipment enables one man to calve cows single-handed in situations where 3 or 4 men would normally be needed. It has a positive, controllable action which is ideal for precise assistance.

A normally-presented calf is born with both front feet appearing first, closely followed by the head. Any departure from this position results in an assisted calving. There are several different ways in which a calf can be incorrectly presented and examination of the cow can identify the problem. Re-adjustment must then be carried out before delivery of the calf can take place safely.

Examination of the cow should be always carried out with care and meticulous attention to cleanliness. The hands and arms should be washed thoroughly in a suitable disinfectant and soaped well, together with the vulva of the cow. This procedure helps to prevent genital infections which could lead to subsequent breeding problems.

Signs of trouble

If a cow stands away from the rest of the herd and shows early signs of calving but nothing appears after about 2 hours, she should be examined. The chances are that there is something wrong. Older cows seem to know if the calf is lying incorrectly for normal presentation and they will give up trying to calve. This can also happen when a set of twins are tangled up with each other. In situations like these, calving the cow should proceed with some urgency as a calf left for too long inside the uterus will die from lack of oxygen.

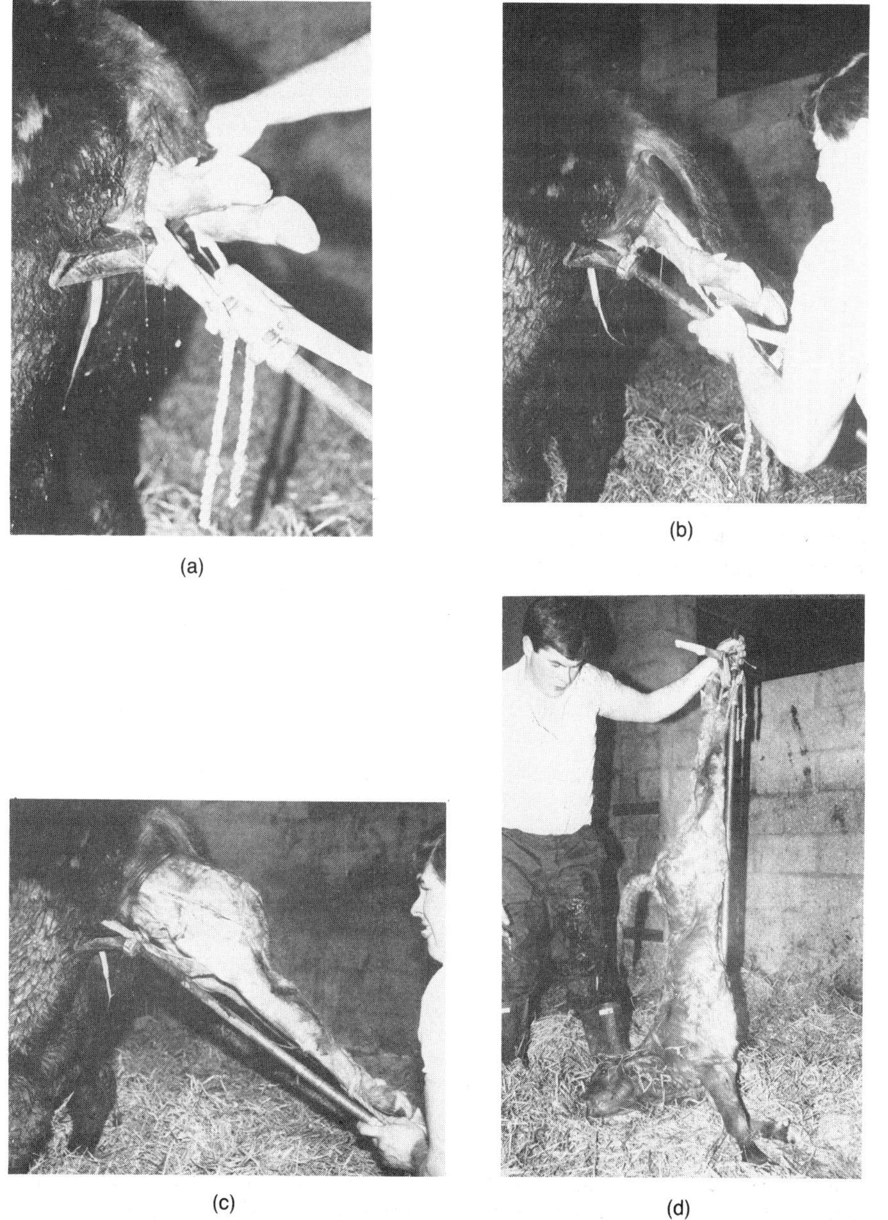

(a)

(b)

(c)

(d)

Calving problems: posterior presentation of calf. Note the use of a calving aid. Immediately after the calf is delivered, it is held up by the hind legs to drain fluid from the mouth, throat and lungs (d).

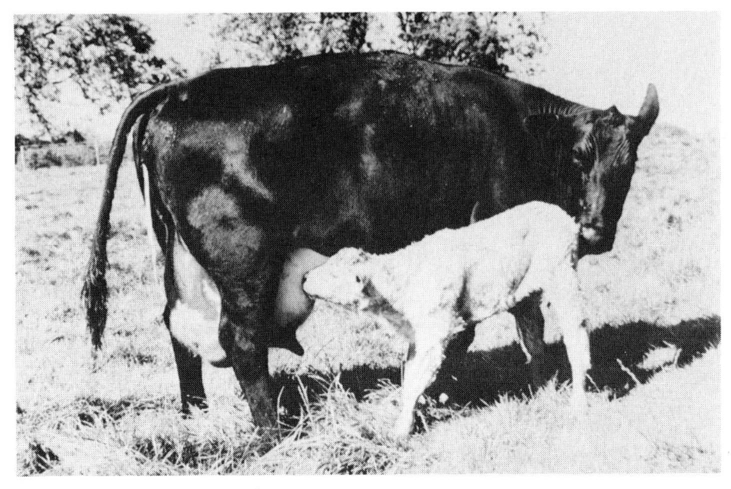

New-born calf in September.

If the front feet have appeared and the cow has been pushing hard for about an hour, but she is obviously making no progress, then she should be caught and given assistance. The head of the calf is probably stuck in the pelvic canal and this should be checked. If it is possible to pass the hand between the head of the calf and the pelvis of the cow then strong positive traction should be sufficient to pull the calf free. If the head locks tight then veterinary assistance should be summoned.

The correct time to apply traction to the calf is when the cow strains with each contraction. The main advantage of a calving aid is that it keeps tension on the calf's legs between contractions, so preventing the calf from slipping back into the cow.

If the back legs of the calf appear first then its hooves point with the soles uppermost and the cow always needs rapid assistance. Strong traction is usually necessary to deliver the calf. As soon as the calf is born it is advisable to hang it up by its hind legs so that any fluid in the air passages can drain out.

Action when calving problems arise

Expertise in calving can only be acquired through experience built up over a period of time, and complicated calvings are best dealt with by a veterinary surgeon until sufficient experience has been gained. It must be remembered that once a cow has started to calve then she should not be left any longer than an hour before she is examined. The longer a cow is allowed to exert useless contractions on a jammed calf then the less chance there is of the calf surviving and the greater is the chance of the cow damaging herself.

CHAPTER 4

CULLING STRATEGY AND HERD REPLACEMENTS

The consistent and long term success of suckled calf production depends on the regular replacement of breeding stock. A specific culling and replacement policy is therefore needed.

CULLING

The basic principle of culling is to remove those cows from the herd which for various reasons are having a detrimental effect on the profit potential of the business.

Reasons for culling

Barren cows

Failure of cows to become in-calf is probably the most common reason for culling. The main cause is usually inadequate feeding prior to and during the mating period and therefore most culling of barren cows could be avoided. However, every year there are likely to be a few repeat breeders for non-nutritional reasons, and these have to be replaced.

Udder problems

These include udders which have been damaged after severe attacks of mastitis. In older cows pendulous udders can result in calves not being able to suckle properly.

Age

It is advisable, as a general rule, to cull cows by about 11 years of age, after they have had 8 calves. As cows get older their milking ability declines, which results in poorer calf performance. Their reproductive performance also declines, which can result in difficult calvings and protracted calving periods, and they are more prone to problems such as milk fever. Another reason for culling cows is that it is generally more difficult to get old cows into good condition for the cull market.

Performance

The performance of suckler cows varies from individual to individual, just as it does in dairy cows. Regular recording of milk production in the dairy herd can identify individual cow performance. In a suckler herd, however, the only means of assessing milking ability is by recording the weight of the calves at weaning. Cows that regularly produce calves below the average calf weaning weight for no obvious reason should be culled.

The sale of cull cows is a valuable source of income and should go a long way towards the costs of replacements. Those cows which are intended for culling should not be put back in calf, and if their condition is relatively poor after weaning then it is well worth feeding them to improve their condition, with the aim of achieving a better sale price.

HERD REPLACEMENTS

In the average situation the target annual replacement rate for a suckler herd should be about 16 percent. This gives cows an average herd life of 7 years.

Source of animals

There are several options open for a herd replacement policy, ranging from buying calves to buying cows with their calves.

Buy heifer calves and rear them

This is the cheapest way of obtaining herd replacements and in many respects it is the most reliable, especially if calves can be bought direct from the farm of origin. About 20 percent more calves should be purchased than are required, to allow for accidental losses, and the disposal of undesirable types and non-breeders. A potential limitation of this method of obtaining herd replacements is lack of grazing and feeding facilities for young heifers within the farm system.

Buy yearlings or bulling heifers

This is a good method because the desired type of heifer can be selected, and control can be exerted over the choice of bull which is used on them. Also, the heifers can be mated to fit in with the calving pattern of the herd. About 10 percent extra heifers should be bought to allow for non-breeders.

Buy in-calf heifers

The main advantage of this expensive method of obtaining herd replacements is quicker calf production, but there is no control of serving date or the type of bull used.

Buy heifers with calves

This is again an expensive method, but at least the heifers have calved safely. However, there is the risk of bringing infection into the herd, especially with cattle bought at a market.

Buy cows with calves

This is also an expensive method of obtaining replacement animals and it is one which is not recommended, due to the uncertainty of the quality and history of this class of stock. The introduction of diseases to the herd is a greater risk than with calved heifers.

The introduction of outside stock into a breeding herd is a risky business because of the challenge that it represents to herd immunity, and the possible introduction of specific diseases. To minimise these risks breeders may be tempted to keep heifer calves which have been bred from their herd to use as replacements. But, as we have seen, this is not advisable because of the declining influence of hybrid vigour.

Angus × Friesian heifer calves; potential herd replacements.

Angus bull running with Friesian heifers.

The best way of minimising the risk of introducing diseases into the herd is to buy replacements when they are at an early age and to mix them with cows from the herd so that they have an opportunity to acquire herd immunity before they calve.

Selection of heifers

Selection of breeding heifers is best done before they are mated, as those which are not considered to be good enough can either be sold as stores or they can be finished for slaughter.

Selection of breeding heifers should be based on:

* Sound conformation
* Frame size
* Good legs and feet
* Steady temperament

Stocky, fat heifers are undesirable and should therefore be culled.

Management of herd replacements

The management of replacement heifers should have the target of achieving a certain weight at the time of mating at 18 months old. The target mating weight varies depending on the breed of cow and Table 4.1 shows the target weights for two popular types of suckler cows.

TABLE 4.1 Target mating weight and growth rate for two types of heifers

	Heifer breed	
	Angus × Friesian	*Hereford × Friesian*
Target mating weight (kg)	340	360
Growth rate, weaning to mating (kg/day)	0.6	0.7

It is important to recognise the need to breed from well-grown heifers because calving smaller animals increases calving problems, and may retard their growth rate and eventually reduce their mature weight. On the other hand, it is extravagant to allow heifers to get too big before they are mated because the investment in the animals does not bring returns soon enough. Young heifers must also not be overfed because their breeding and milking ability can be adversely affected by excessive fatness.

It is sound management practice to serve heifers so that they start to calve 4 to 6 weeks before the main herd. This allows more time for recovery after calving and for regular oestrus to get underway. Heifers generally have small calves, so calving early in the season allows more time for them to reach better weights by the time they are weaned. The choice of the bull that is used on heifers must be based on calving ability because heifers are prone to calving difficulties, the incidence of which should be minimised as far as possible. The use of native breeds is the safest course to take, but of the continental breeds the Limousin and Blonde d'Aquitaine are probably the safest breeds of sire to use.

After calving, in a housed situation heifers should be separated from the rest of the herd to allow better control of feeding. They require extra feed in a non-competitive environment to allow for growth as well as for lactation. It may well be necessary to wean the heifers before the cows, especially in a poor grass year, in order to give preferential feeding to improve their condition before they calve again.

It is important to remember that heifers must be looked after with care during their first lactation, to ensure that they are in good enough condition for rebreeding to take place successfully. Failure to achieve this will either waste valuable time and resources or may result in the heifer having to be culled, with the subsequent loss of a potentially productive animal.

CHAPTER 5

THE HERD SIRE

The herd sire has a major effect on the profitability of suckled calf production. The old saying "the bull is half the herd" is an understatement of the real influence that a bull exerts on herd performance. Although the dominant features that a bull passes on to his progeny are growth rate and conformation, he also has considerable influence over calving performance.

CHOICE OF SIRE BREED

The choice of the breed of sire to be used on a herd depends on several criteria including:

Calving difficulties

The heavier the sire breed the more likely the incidence of calving difficulties, and calf mortality rates are likely to increase. The choice of breed should therefore reflect the level of commitment to monitor closely the herd during the calving period. At the same time, however, calving difficulties can be considerably reduced by correct cow management and by selecting individual bulls which are likely to give fewer calving difficulties.

Supervision during calving is desirable in all herds to ensure the maximum number of calves born alive. But particular care taken where, say, a Charolais bull is used is very well rewarded by the extra value of the calf crop.

Breed of dam

The need to use bulls of a different breed to that of the cow has already been stressed (Chapter 1).

Calf disposal

The heavy breeds of cattle, such as the Charolais and Simmental, produce progeny that finish for slaughter at higher weights; they also provide leaner carcases than do our native breeds. Consequently, if the calves are sold as stores they command premium prices, and if they are retained for finishing then the added value of the carcases is of great benefit to the farm business. The effect of sire breed on calf weaning weight is shown in Table 5.1.

TABLE 5.1 Effect of sire breed on calf weaning weight

		Type of herd	
	Lowland	*Upland*	*Hill*
Sire breed		*200-day weight (kg)*	
Hereford	208	194	184
		Difference from Hereford cross (kg)	
Charolais	+32	+33	+21
Simmental	+24	+28	+14
South Devon	+23	+27	+16
North Devon	+17	+21	+ 7
Lincoln Red	+14	+20	+ 5
Sussex	+ 7	+13	+ 2
Limousin	+ 7	+10	+ 2
Aberdeen Angus	−14	−12	− 8

Source: Meat and Livestock Commission

SELECTION OF INDIVIDUAL BULLS

Selection of the individual bull must be taken very seriously because the final choice has a profound effect on the profitability of the herd. There are great variations of type and size within each breed, so it is advisable to learn something of the proven bloodlines of the chosen breed to help in the final selection. Sons of cattle which have outstanding breeding records behind them are obviously desirable.

It is important to gather as much information as possible about intended herd sires. It is not good enough to try to judge them by eye alone. Equally, records alone may be deceiving because individual bulls may have been overfed and pampered to achieve inflated recorded weights. Overfeeding can have adverse effects on foot development and on fertility, and weight gains taken in isolation do not indicate the likely breeding value of a bull.

The ideal bull is one whose progeny are born easily, grow rapidly and produce carcases of good conformation. Because these criteria are inherited the MLC has devised the Beef Selection Index which indicates with a single score the overall breeding assessment of individual bulls. The method used to arrive at an appropriate index for each bull involves group performance testing and it incorporates ratings for calving difficulty, birth weight, 400 day weight, ultrasonic backfat measurement and muscle score. The indices range from 50 for poor bulls to 150 for the most outstanding bulls. Breeders should look for bulls with an index of 120 or more.

Goldies Tarzan: a Charolais stock bull used with the Givendale herd.

When a bull is considered which has not been scored by this index, it is important to find out his birthweight and whether he, himself, was born without difficulty. His weight record should be studied and his 400 or 500 day weight should be above the breed average shown in Table 5.2. It must nevertheless be remembered that the most reliable information is that from MLC group performance testing.

TABLE 5.2 Average weights recorded 1985 to 1987 for different breeds of beef bulls

	Average weight (kg)	
	400 days	*500 days*
Charolais	604	729
Simmental	581	694
South Devon	551	654
Limousin	530	635
Blonde d'Aquitaine	515	634
Belgian Blue	521	610
Hereford	459	551
Aberdeen Angus	441	546
Lincoln Red	485	NA
Sussex	446	544

NA Not available
Source: Meat and Livestock Commission

Having studied all the relevant information which is available, the final selection must be made with the physical soundness of the bull in mind. Desirable physical features include:

* Sound legs and feet, combined with good action
* Clean shoulders and a good, level topline
* Well-fleshed and good muscling over the rump
* Minimum brisket and belly
* Good length
* Masculine head and appearance
* Steady temperament.

Undesirable points which should be avoided are:

* Straight or sickle-hocked hind legs
* Bulls with small testicles, as these can lead to fertility problems
* Young bulls which carry a lot of fat or belly because they have probably been overfed and will quickly 'melt' when put on a commercial breeding regime. Fat bulls are not a sound long term investment.

MANAGEMENT OF BULLS

Considerable capital is invested in a bull which has the desired characteristics, and to get the best out of this investment he must be managed with care and attention. If a bull is properly looked after he can still serve cows successfully when he is 7 or 8 years old. However, the average working life of a bull is much shorter than this, which in many cases is due to inadequate care of the animal.

It is important to keep the herd sire in good physical condition, not only to ensure a long working life to justify his purchase price, but also because the condition of a bull prior to a working session has a direct bearing on the conception rate that he is likely to achieve. He must be physically strong but not fat, his feet must be in good condition, he must be fertile and he must have the desire and the ability to deposit his semen in the cow.

Nutrition

The correct feeding of the young bull is very important to ensure his continued growth and proper development. A bull is usually purchased when he is 15 to 18 months of age, and if the right bull has been selected then he will still be growing at about 1.5 kg per day. This growth rate should be maintained by feeding him about 9 kg a day of a mineralised cereal and sugar beet pulp ration of about 14 percent crude protein. The growth rate and quantity of feed suggested refer to large continental breeds; smaller British beef breeds will have a lower growth rate and will require less feed.

The age at which bulls mature depends on the breed, with bulls of larger breeds maturing later than bulls of smaller breeds. As a general rule, the larger breeds grow rapidly until they are about 2.5 years of age but their final mature size is not reached until they are about 4 years old. Bulls over 2.5 years old require less cereal-based feed because they are able to utilise predominantly forage-based diets to sustain a declining growth rate.

Exercise

In addition to adequate nutrition, exercise is important when the bull is laid off work. A small, well-fenced paddock is useful for this purpose during the summer, and in the winter a bull pen with sufficient exercise room should be provided. Before constructing a bull pen, advice on the legal requirements concerning bull penning and handling should be sought from the Health and Safety Executive.

Foot care

The majority of bulls have sound feet which require only minor trimming periodically, but unfortunately there are some bulls which need frequent attention to ensure their continued mobility. Correct foot care will prolong the working life of any animal, but foot trimming is a particularly neglected task where bulls are concerned because they are difficult to handle without the right facilities. All too often nothing is done until the bull goes lame, which often occurs when he is working, possibly due to the extra strain that he exerts on his feet when mounting the cows. The bull may then have to be rested to allow time for the injured foot to heal, and a serious financial loss may be incurred because another bull may have to be hired or bought in a hurry, in order to get the cows in calf on time.

Foot injuries can be caused by foreign objects penetrating the sole of the hoof. I have seen, to my horror, both bulls and cows in fields and farmyards which are littered with scrap iron and old posts with nails sticking out of them. It is amazing that the owners seem to be unaware of the dangers that these objects pose to their valuable livestock. A deep, penetrating wound into a hoof can cause permanent damage and the eventual untimely slaughter of an animal, invariably accompanied by veterinary costs.

Routine foot trimming should be considered essential, but the length of time between treatments depends on how quickly the feet of an individual bull deteriorate. Generally speaking, all bulls should have their feet checked once a year, but more frequent inspections may well be necessary for bulls with a history of bad feet.

Foot trimming a bull in a home-made crush.

There are two ways of approaching foot trimming. Firstly, the bull can be sedated by a veterinary surgeon who then trims his feet while he is laid out. Secondly, a foot trimming crush can either be bought or made on the farm and the job can then be tackled by a competent stockman who has received adequate training. Generally speaking, foot trimming is not a difficult task and it is a skill which is well worth learning because routine work can be undertaken regularly without the worry of escalating veterinary costs.

Assessment of fertility

It is of course essential for a bull in a commercial situation, serving a lot of cows quickly, to have good sperm production. A good guide to the likely fertility of a bull is the size and shape of his testicles. They should be well developed, feel firm and be of equal size.

Mating records should be kept in order to verify a bull's conception rate. It is also advisable to check by observation that a bull continues to serve his cows properly because it is not unknown for a bull to damage his penis in such a way as to make serving impossible.

As a bull gets older the quality of his semen tends to deteriorate, and it is therefore well worth having it checked annually once a bull reaches about 6 years old. This can be done on the farm by a Milk Marketing Board veterinary surgeon.

Working a young bull

Young bulls must not be expected to get a lot of cows in calf when they are at an early age. Age at sexual maturity varies depending on the breed and it can also vary between individuals of the same breed. Generally speaking, the bigger the breed then the later sexual maturity is reached.

Initially a young bull should be mated with a cow which is in standing heat, away from other stock. Once he has been seen to serve the cow then he can be turned into a herd of no more than 20 cows. It is a bad policy to run a young bull with a larger number of cows because poor conception rates could result, giving a protracted calving period. Larger numbers of cows could also result in the bull losing too much condition and even suffering growth retardation.

It is important to keep a careful watch on a young bull to make sure that he continues to work properly. It is also advisable to take him away from the herd after 3 weeks and to replace him with an old bull. This strategy covers for the young bull in case he is sub-fertile or infertile, and keeping a check on the cows which return for re-mating provides valuable information on his success.

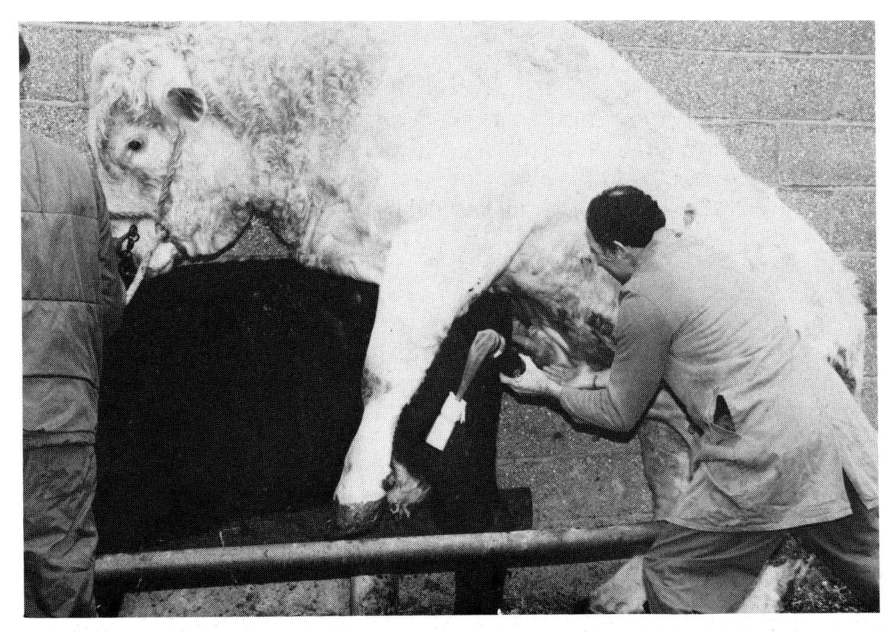

Semen collection for examination to test fertility.

If the first working session of a young bull occurs during the winter, special care must be taken to ensure that he is adequately fed and it may be necessary to feed him individually. In a confined area bullying from older cows can be a problem. There is a danger of chills developing after sweating. To prevent injury to the bull it is important to ensure adequate bedding and to avoid slippery surfaces such as those found in cubicle sheds.

PROGENY TESTING

The progeny testing work that the Milk Marketing Board (MMB) undertakes show quite marked differences in the performances of individual bulls of the same breed. The MMB is looking for bulls which are best at producing calves which are born easily, grow rapidly and produce carcases with a high yield of lean meat. These are exactly the same criteria that suckled calf producers should be looking for, but how can bulls which are superior at doing the job be identified in a farm situation?

In larger suckler herds where two or more bulls are used there is the opportunity to set up a comparative progeny testing scheme to verify the breeding merits of individual bulls. The results of progeny testing not only measure the breeding value of a bull, but they also highlight how well the bull was selected in the first instance.

A group of young Charolais bulls during on-farm performance testing.

Progeny testing involves recording data about every calf sired by the bull. Information recorded should include:

* Parentage
* Calf birthweight and calving score
* Weaning weight
* Growth rate
* Calf price if sold at this stage.

Data from calves sired by different bulls can then be compared. If the calves are finished for slaughter on the farm then the comparisons can continue. The following further information is required:

* Growth rate to slaughter
* Age to slaughter
* If sold deadweight—carcase appraisal
* If sold liveweight—price per kilogram
* Final value.

Progeny testing is a relatively slow method of testing bulls but it confirms the breeding value of individuals, and this information can be used to cull bulls which give serious calving difficulties or inferior calves. On the positive side, it highlights the true value of a superior bull, which can be measured by the extra income generated by the increased sale value of his calves.

Progeny testing has the added advantage of providing information on bulls from which semen may be sold. It is also very useful to be able to relate commercial progeny testing to a pedigree breeding programme.

CHAPTER 6

MANAGING THE SUCKLED CALF

For a calf to achieve its full potential it must have adequate milk from its mother. It must also have access to creep feed at certain times of its life. The health status of a calf also has a considerable bearing on how quickly it grows and on the accumulation of labour and veterinary costs. Considerable benefits both in terms of growth and conformation can be gained by leaving the bull calves entire, although adjustments to management practices may be necessary.

NUTRITION

Colostrum

The future well-being of a calf depends on the early absorption of sufficient quantities of colostrum. A calf is born without any antibodies to protect it from bacterial infections such as septicaemia, joint ill, pneumonia or scour. The antibodies which are acquired by the calf for this protection are carried in the colostrum, which is also high in nutrients and vitamins.

The ability of the calf to absorb antibodies decreases after 6 hours from birth, so it is essential to observe that a newly born calf has suckled its mother within this period. Special care should be taken with first-calved heifers to make sure that they mother their calves and that they actually allow their calves to suckle. Where a calf is unable to suckle its mother because of injury or some other circumstance, the cow must be milked and 1.5 litres of colostrum given to the calf for its first feed. In this situation it is unwise to try to force the calf to drink, so the colostrum should be given with the aid of a stomach tube to avoid waste and to avoid passing fluid into the lungs of the calf.

There are no substitutes for colostrum so it is advisable to have a deep frozen supply of colostrum from a cow which has calved previously, to cover for emergency situations when colostrum cannot be provided by the calf's dam.

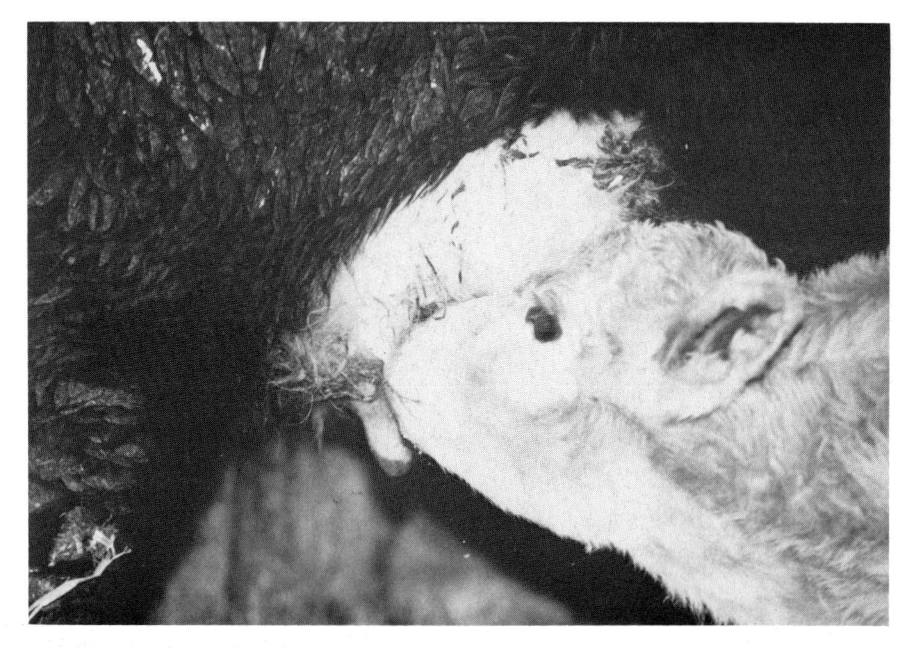

New born calf sucking colostrum.

Milk

A good supply of milk is essential for calves to continue to grow to their full potential. The advantages of hybrid vigour in this respect have already been discussed (Chapter 1), and the nutritional needs of cows to ensure adequate milk production were covered in Chapter 3.

Creep feeding

Creep feeding has a beneficial effect on growth rate at times when the nutritional value of milk and forages is insufficient to maintain the growth potential of the calves. As calves grow older the effect of milk on growth rate is reduced because, with the development of the rumen, emphasis moves from milk to forage as the major source of nutrients. The creep feeding of calves should therefore be considered in situations when milk yield is reduced due either to a restriction in the amount of feed offered to cows or to a decline in the nutritional value of grass. The most effective way to utilise creep feed is to anticipate occasions when the calves are likely to gain the most benefit, and then to increase intake slowly during these situations.

Area for creep feeding autumn-born calves. The vertical bars are too narrow for cows to enter the area, but they are wide enough apart for calves to pass through to eat the creep feed.

CREEP FEEDING STRATEGIES

Autumn born calves

The creep feeding of autumn born calves is well worthwhile especially during February, March and April because milk production from the cows declines during this period. The reduction in milk yield comes about not only because of advancing lactation but also because savings may be made in the feed of cows at this stage of their reproductive cycle. It is more efficient to feed cereals directly to the calf at this time than it is to feed the cow for lactation.

Autumn-born calves start to take small quantities of creep feed by the end of November. As they grow older intake increases, until by the end of February the calves may be eating 2 kg per head per day. By this stage the rumen has developed so it is also beneficial to give the calves good quality silage or hay. Provision of a separate area where the calves can lie away from the cows and where creep feed and forage can be fed is necessary.

Autumn born calves (300 kg) creep feeding in April. Creep feeding gives an advantage in terms of weight gain.

When the herd is turned out in the spring, creep feed is not needed because the calves benefit from a good supply of nutritious grass, in addition to a flush of milk from the cows. Later in the grazing season, when the quality of grass and the supply of milk declines, it may be desirable to feed the calves a cereal supplement. This is a time-consuming task because the calves have to be separated from the cows every day to be trough fed, although after an initial training period it is not as difficult as might be imagined.

The advantages of this sort of supplementation are apparent when:

a. The calves are housed for finishing because the change in ration is less abrupt, or

b. The calves are sold, especially at suckled calf sales, as they have 'bloom' and have gained extra weight.

Winter born calves

There are no advantages in creep feeding calves which are born after January because the supply of milk from their dams should be adequate to carry them through to turn out. After the herd is turned out in the spring, the calves are able to take advantage of the boost in milk production.

February-born bull calves using creep feeder in June.

Creep feeding can, however, make a significant contribution towards calf growth from July onwards, and creep feeders should be put out during June so that the calves get used to going into them. In order to ensure that all the calves have access to the ration, one 5-space feeder should be available for every 15 to 20 calves.

Initially, small amounts of creep are taken but as grass growth and milk supply declines during July, the intake of creep feed increases. This pattern continues during the late summer, until the calves are weaned in October. Calves which are fed in this way not only weigh considerably more when they are weaned but they also take more readily to concentrate feeding when they are housed, so reducing stress and weight loss at this time.

Spring born calves

Calves which are born in May and June benefit from creep feeding in the autumn for the same reasons that have been indicated for winter born calves. The method of feeding should be exactly the same, but because the calves are younger creep feeding need not start until the end of August.

Disbudding a calf in a home-made calf crush designed to reduce stress.

Weighing and worming a calf at weaning.

CALF HEALTH

Diseases in suckled calves must be avoided as far as possible to ensure that the calves grow to their full potential. Correct management can go a long way towards minimising the risk of diseases, but when problems occur then it is essential that treatment is carried out swiftly and positively. It is important to anticipate the likely situations which can lead to calves developing diseases so that preventative measures can be taken.

The essential role that colostrum plays in building-up resistance to bacterial infections of young calves has already been considered. Stressful situations associated with either handling calves or with adverse environmental conditions can trigger off diseases such as pneumonia or scouring, but the incidence of disease can be reduced by good management, as follows:

* Handling should be kept to a minimum, and routine procedures such as dis-budding and castrating should be done in favourable weather conditions to reduce stress.

* Buildings should be well ventilated and calves should have a dry, draught-free lying area which must be kept well bedded.

* Weather conditions are of particular significance where respiratory diseases are concerned. Spring and winter-born calves are vulnerable to adverse weather so a careful watch should be kept on calves during cold, windy and wet spells. Housed calves in the winter are at risk from outbreaks of pneumonia, which seem to be directly dependent on the prevailing conditions. The higher the atmospheric humidity, the greater the number of airborne pathogens and the greater the risk of an outbreak of pneumonia. It is therefore important to be extra vigilant during foggy conditions or during a thaw after a cold spell.

* One of the risks associated with calving suckler cows in straw bedded yards is the problem of scouring calves. This situation can largely be overcome by vaccinating the cows before they calve. It is important to remember that scouring calves must be treated very quickly, with veterinary advice. Navel infections and joint-ill can also be a problem if navels are not treated soon after the calf is born.

* Deficiency of certain trace elements such as copper or selenium can be a problem because they are essential nutrients for calves and cows. Veterinary advice should be sought if, for no apparent reason, a proportion of animals in a herd are unthrifty.

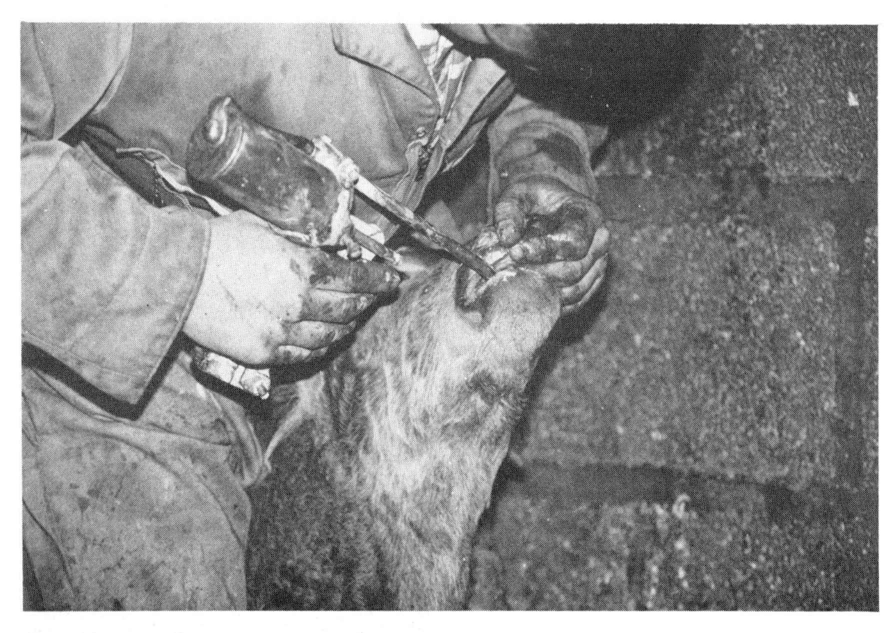

Drenching a calf.

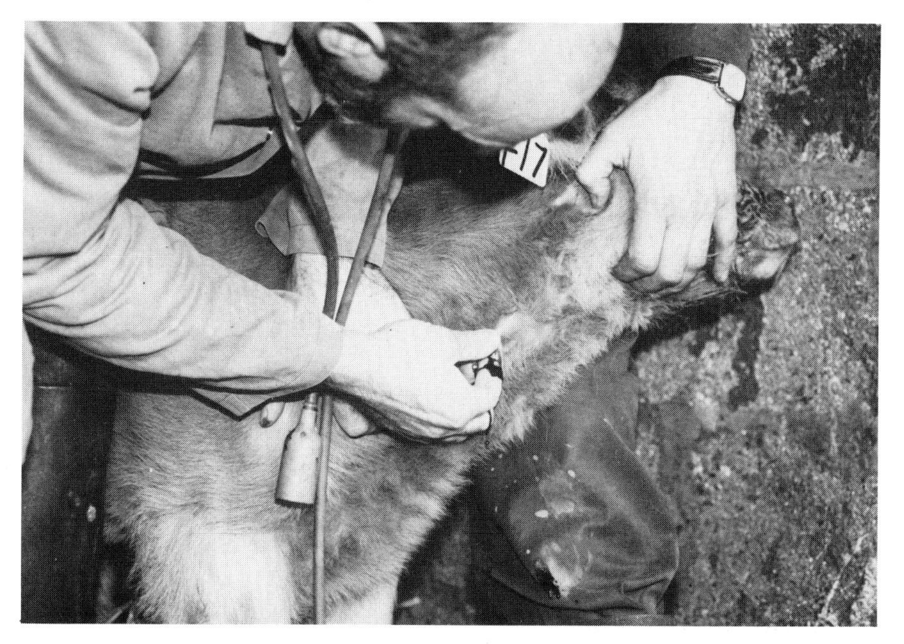

Intravenous saline treatment to correct dehydration in a scouring calf.

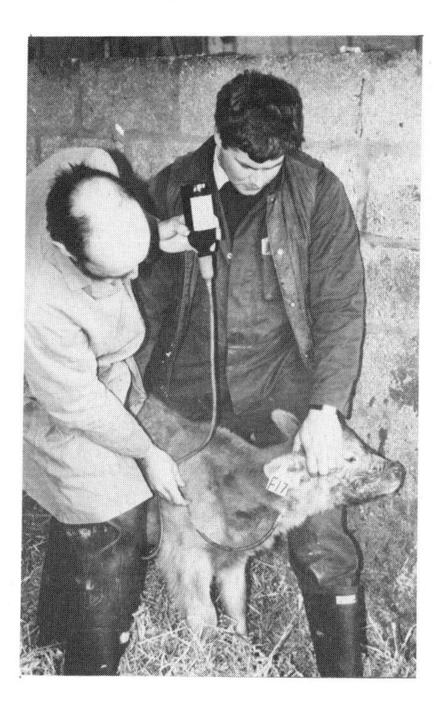

Stomach tubing a calf with electrolytes as part of the treatment for scour.

LEAVING BULL CALVES ENTIRE

The fact that bulls perform more efficiently than steers provides an opportunity to increase the value of male calves. Young bulls are becoming more popular with butchers because of their superiority over steers with regard to leaness and to yield of saleable meat. The evidence now available suggests that demand for bull beef will increase, although the production of bull calves in the more traditional suckler areas may be resisted in the short term.

The following points must be considered when bull calves are left entire:

* Cows with bull calves must be split from those with heifer calves when the calves are about six months old. This ensures that young bulls cannot harass heifers.

* Separate, well-fenced grazing areas must be available for calves older than about six months.

* Cows that have twins of mixed sexes must have their bull calf castrated so the family group can run with the heifer calf group.

* Cull cows which have bull calves at foot and which have not been put back in calf must run with the heifer group, with their calves castrated. Cows which come bulling in a group which is suckling bull calves can cause considerable disturbance and stress, therefore it is essential for cows to be pregnancy-diagnosed in good time so that only pregnant cows run with this group.

CHAPTER 7

RECORDING HERD PERFORMANCE

THE NEED FOR RECORDS

The importance of developing well-organised management standards to achieve profitable production has already been emphasised. Recording plays a vital role both in the effective day-to-day supervision and in the longer term planning of a suckler herd.

Records clearly have an important part to play in the preparation of costings. In addition, through the actual process of recording attention is focussed on the performance of the herd, and this stimulates a desire to improve and to refine management techniques.

The main elements which need recording are:

1. *Feed*: accurate records of the type and quantity of concentrate and forage fed.

2. *Grassland*: records of grazing and conservation areas to evaluate grassland utilisation.

3. *Calf performance*: calf growth rate calculated from calf weights recorded at different periods of the year, such as at birth, at turnout, at weaning and at housing.

4. *Breeding*: records of mating and calving details for individual cows, to assess the performance of both dams and sires.

5. *Financial*: records of the purchases and sale of stock and the cost of other items, such as veterinary treatment.

KEEPING ON TARGET

The first step is to set production targets. The recording system may then be used to ensure that the targets have been met, so keeping the herd on course for profitable production.

The MLC has gained a wealth of experience in suckler herd management over many years and the Beefplan service which it offers tailors production targets to suit individual farms. The average production targets published by the MLC for suckler herds in 1986 are shown in Table 7.1.

TABLE 7.1 Performance targets for suckler herds

	Continental sire[1]		*British sire*[2]	
	Autumn calving	*Spring calving*	*Autumn calving*	*Spring calving*
		Lowland herds		
Calving period (max. days)	90	85	90	85
Calves weaned/100 cows bulled	92	92	95	95
Calf daily gain (kg)	1.0	1.1	0.9	1.0
Sale weight of calf (kg)	350	280	300	230
Stocking rate (cows/ha)	1.9	2.3	2.0	2.4
		Upland herds		
Calving period (max. days)	90	85	90	85
Calves weaned/100 cows bulled	92	92	95	95
Calf daily gain (kg)	1.0	1.1	0.9	1.0
Sale weight of calf (kg)	350	280	300	230
Stocking rate (cows/ha)	1.5	1.9	1.6	2.0
		Hill herds		
Calving period (max. days)	100	90	100	90
Calves weaned/100 cows bulled	85	85	90	90
Calf daily gain (kg)	0.9	1.0	0.8	0.9
Sale weight of calf (kg)	310	265	260	220

(1) Eg Charolais
(2) Eg Hereford

Source: Meat and Livestock Commission

HERD RECORDS

Herd records play a very important part in the day-to-day running of the herd. Cow and calf identification forms the basis for recording essential information which can then be used when making management decisions.

Individual identification is important for:

1) Breeding records
2) Resolving health problems
3) Determining the age of cows
4) Matching cows with their own calves

Every herd in the country has its own Ministry approved number which is stamped on ear tags which must be used on every calf born in the herd. The first step in setting up an individual recording system is to implement, in addition, a *cow* numbering index. This is best achieved with the use of numbered plastic ear tags. A record must be kept of both the *herd* tag number and the *cow* tag number as plastic tags may be accidentally torn out of the ear, leaving the herd tag as the only means of identification. The age of cows can be expressed by using a letter prefix on the plastic tag which relates to the year of entry into the herd. Cows can then be easily dated at a glance, which proves useful when making management decisions.

The ear tags of the whole herd should be checked twice a year and any that are found to be missing or illegible should be replaced. It is essential to carry out one of these checks prior to calving because every calf should be tagged to correspond with the number of its dam.

BREEDING RECORDS

As we have seen, increased production from a suckler herd can be achieved by selecting breeding stock which have superior natural growth potential. Recording plays an essential role in this selection process, as shown in Figure 7.1.

FIGURE 7.1 Relationship between recording and selection of breeding stock

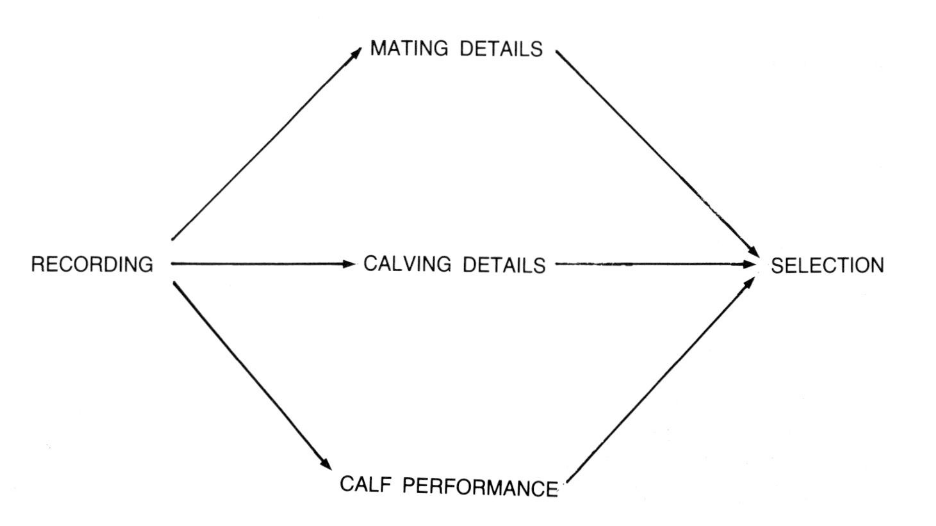

Both cow and bull performance are so interrelated that a full recording programme is worthwhile in order to present a detailed summary of the breeding stock involved. Records should include information under the following headings:

Dam Ear Number
Sire Code Number
Calving Date
Calving Code (denoting ease of calving)
Birth Weight
Calf Ear Number
Calf Sex
Weaning Weight
Weaning Date
Sale Date
Sale Weight
Sale Price

It can be seen from these headings that a full picture can easily be built up of the breeding values of individual cows and bulls. Where 2 or more herd sires are used comparative progeny evaluation of each bull can be carried out, bringing factual accuracy into sire selection. This system of recording also identifies those cows which need culling for consistently producing inferior calves.

CHAPTER 8

GRASSLAND MANAGEMENT AND FORAGE CONSERVATION

It is important to recognise the need to achieve efficient grassland performance, and to plan the grazing and conservation requirements of the herd with care. Stocking rates are significantly influenced by the effect of fertiliser on grass production, and this effect can be exploited to increase the size of the suckler herd. The majority of lowland suckler herds graze marginal land which is unsuitable for cultivation—land which is either steep, stoney, badly drained or parkland. But the need for conserved winter feed must be met, and where silage has been selected as the main forage the utilisation of temporary leys may be necessary. Silage may also be taken from areas of permanent grassland which can then be grazed later in the season.

PERMANENT GRASSLAND

Managing a permanent grass grazing system is a delicate balancing act between grass supply and cow demand. The supply part of the equation involves encouraging the sward to produce a high output of grass, while on the other hand ensuring that stocking rate is at an adequate level to make full use of the grass produced. Cow demand is largely governed by the size of the animal and the stage of her production cycle.

The main obstacle to achieving the balance is variation in grass production from season to season, due mainly to fluctuations in rainfall and in temperature. Good sward management involves a number of factors and failure in any one will have a detrimental effect on final grass production, as illustrated in Figure 8.1.

Freedom from weeds

The sight of grazed pastures which are knee high in numerous weeds all doing their best to choke out the productive grasses in the sward is all too common. It is difficult to understand why farmers allow this situation to occur when they go to great lengths and expense to keep their arable crops free from weeds—after all, grass is a crop. But traditionally it has been mis-understood, mis-treated and therefore mis-managed.

FIGURE 8.1 Factors contributing to a productive sward

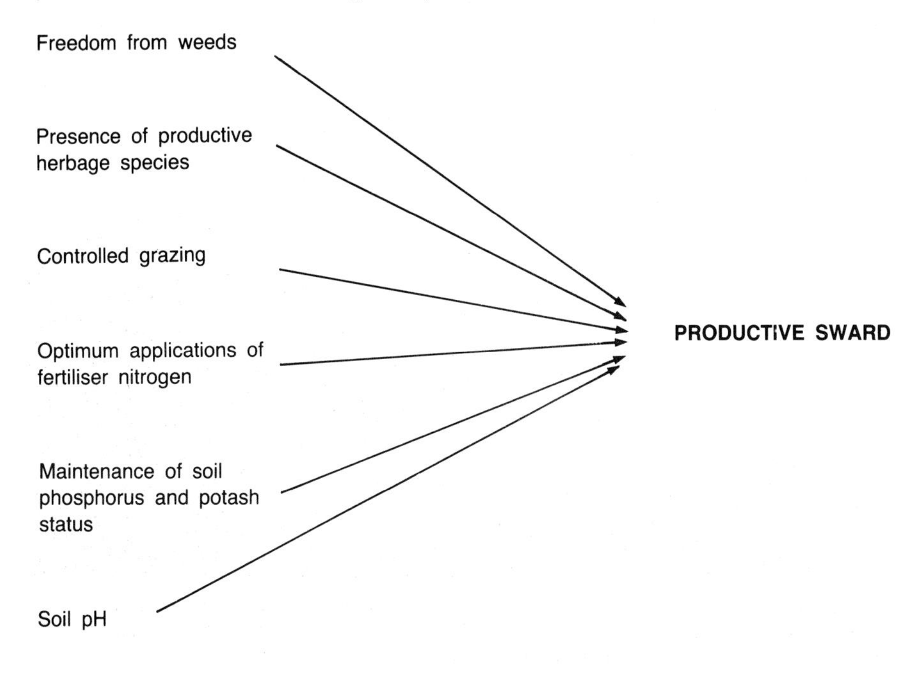

The removal of weeds is as essential in a grass crop as it is in an arable crop. Several selective herbicides are available but care must be taken not to damage any clover which may be present in the sward.

Sward type

In a situation where grassland is difficult to re-seed, for example on a steep hillside, encouragement of indigenous grasses is probably the best course to take. If, however, the land is ploughable then it may be worthwhile to undertake a re-seeding programme using a perennial ryegrass and white clover mixture to improve the pasture.

Clover can be an important component of the sward, and it has a major contribution to make in suckler cow and sheep enterprises. The 'D' value of clover varies little during the grazing season and trials have shown that liveweight gain and milk yield are improved with a higher intake of clover. Its inconsistency in adverse weather conditions, however, leads to disappointing nitrogen fixation in some seasons.

Top dressing permanent grass by air in early March. It is important to apply fertiliser early in the season, once T sum 200 has been reached, but in the situation illustrated ground conditions are too wet and dangerous in March to apply fertiliser using a tractor.

A 4-wheeled drive tractor top-dressing permanent grassland at the end of May.

Controlled grazing

It has been shown recently that sward height has a considerable influence on the utilisation of grass and on animal performance. The established guidelines for grazing swards indicate that the average height of the sward should be maintained between 6 and 10 cm. The shorter height is appropriate in the early part of the grazing season, while taller swards are required in the autumn.

Sward height is used as an indicator of the likely nutritional value of the grass. The most nutritional grass is young, fresh growth, which has a high 'D' value. If a sward is understocked, grass height gets out of control and the cattle graze only parts of the area. In this situation the undergrazed areas soon run to seed and are wasted.

Controlled grazing therefore involves manipulating stocking rate to prevent plants going to seed. This can be achieved by setting up a series of paddocks which are grazed in rotation or by set stocking with the correct number of animals. The latter option is in many ways the better of the two, although it is the most skillful to perfect and it relies on past experience of the potential of the available grassland. Certainly, cattle seem to be more settled and therefore 'do' better in a set-stocked situation. If grass growth does get out of control and seed heads appear, the pasture must be "topped" to encourage fresh growth.

Fertiliser nitrogen

Grass growth in early spring is influenced by temperature. A technique developed in Holland has resulted in the widespread use in the UK of a method, known as the T sum, to decide when to apply the first dressing of nitrogen fertiliser. The T sum is calculated by adding all the positive daily air temperatures from January until the accumulated total of 200°C is reached. This is known as 'T sum 200' and nitrogen should then be applied as soon as conditions allow. In most areas the T sum 200 is reached by mid-March, well before the more traditional April time of first application.

Surveys conducted by the Agricultural Development and Advisory Service of the Ministry of Agriculture (ADAS) and by the Institute for Grassland and Animal Production (IGAP) have shown that grazing land has an advantage of increased output when nitrogen is first applied at T sum 200.

The total amount of nitrogen applied in individual cases depends on the intended intensity of stocking. A higher stocking rate obviously requires more nitrogen than a lower stocking rate. Past experience of the capability of a particular area of land to grow grass with increasing nitrogen levels can result in a steady rise in stocking rate.

Cows grazing well-managed pasture. Controlled grazing is practised to maintain grass height at the optimum for pasture utilisation and for animal performance.

Soil pH and phosphate and potash status

Soil analysis can determine the status of soil acidity (measured as pH), and phosphate (P) and potash (K) levels. For good grass growth the soil pH must be 5.8 or above, and a soil index of 2 for both potash and phosphate are required—any deficiencies should be corrected by liming or applying the correct fertiliser.

BUFFER FEEDING

In situations where stocking rates are high, there is always a possibility that grass will run short in adverse weather conditions such as drought or unseasonally low temperatures. If a grass shortage does occur then it may be necessary to give the herd extra forage—straw, hay or silage—in order to meet the nutritional requirements of the cows and calves. This is termed buffer feeding.

Buffer feeding the winter calving herd with straw in September.

GRASS CONSERVATION

The conservation of grass is a vast subject which is covered in many publications. This section therefore is limited to a brief look at the basic principals of silage quality and how different forms of conservation may be utilised in suckled calf production.

In the majority of cases conservation needs are probably best met by the utilisation of temporary leys. Modern Italian ryegrasses are ideal for silage making, giving high yields of material with a good feed value. A temporary ley also makes an ideal arable break crop and is useful for grazing weaned calves after the silage cuts have been taken. In situations where permanent grassland is flat enough for forage equipment to work safely, it may be possible to capitalise on spring grass by taking an early cut of grass on part of the area. Grazing can then be extended onto the conservation area as grass growth declines during mid-summer.

Contrary to the opinions of many, I believe that silage *quality* is very important where the feeding of suckler cows is concerned. I have already emphasised the relationship between feeding, body condition and reproductive performance (Chapter 3). The success and profitability of herds that calve in the autumn and winter hinge on good quality silage. Spring calving herds on the other hand only require a maintenance level of feeding during the winter so silage with a lower 'D' value and a higher fibre content is adequate—the advantage here is that a bigger crop of material can be taken when the grass is more mature, so increasing stocking rates.

Forage harvesting grass for silage.

The silages which I aim to feed to the autumn and winter calving herds have the target analysis shown in Table 8.2:

TABLE 8.2 Target analysis for silage

	Target
pH	4.1
DM (%)	25
CP (% in DM)	16
DCP (g/kg DM)	105
MAD fibre (% in DM)	28
ME (MJ/kg DM)	10.8
D value	67
Ammonia N (% total N)	8
Ash (% in DM)	9.0

Silage of this quality requires minimal cereal supplementation to achieve the desired cow condition for good reproductive performance and for adequate milk production. Poorer quality silage requires additional feeding of cereals to achieve the same results, which inevitably increases costs.

The silage making system at Givendale is based on temporary leys because the permanent grassland is too steep to cut. The grass is cut when young, just before the seed heads appear, leaving a 3 inch stubble to avoid soil contamination. A drum mower/brush conditioner is used and the swath is left as wide as possible to increase wilting rates. The grass is picked up with a multi-chop machine usually 24 hours later, depending on weather conditions. An acid-based additive is used to ensure a rapid reduction in pH in the clamp, which is filled as rapidly as possible and rolled well before sheeting.

Big-bale silage is becoming more popular as bale wrapping techniques become more reliable. On many farms big bale silage has replaced hay, because it is less dependent on dry weather during the summer.

This method of conservation does not require expensive silage clamps and it is often carried out by contractors. It is particularly useful where relatively small quantities of silage are made, where there is an opportunity to conserve grass surplus to requirements later in the season, and where transport of silage from a clamp would be difficult or labour-intensive.

CHAPTER 9

THE SUCKLER HERDS AT GIVENDALE FARM

BACKGROUND

Givendale Farm was acquired in 1974 by JSR Farms Ltd. The farm had been badly neglected so it was necessary to devise a sound management plan which would fit the resources of the farm and which would also achieve a worthwhile return on tenants' capital. The farm plan was dictated by the fact that one third of the area was steep, unploughable grassland. It was also recognised that the poor quality arable land needed to be improved if cereal yields were to reach profitable levels.

The financial viability of Givendale clearly hinged on the development of a system which would result in the successful integration of both livestock and arable enterprises. So the decision was taken to utilise the permanent grassland with suckler cows and a breeding ewe flock. Winter forage for both enterprises would be produced from straw and 2-year leys, grown as a break crop in the arable rotation.

The farm, which extends to 402 hectares, is situated on the western edge of the arable Yorkshire Wolds. The arable land is cropped with cereals and grows 2-year leys. It comprises stoney, chalk wold, which is mainly sloping and exposed, rising 213 metres above sea level. The unit is now split into 271 hectares of productive arable land and 131 hectares of unploughable permanent grassland and steep dale. A plan of the farm is given in Figure 9.1, with a profile showing the topography in relation to land use in Figure 9.2.

After considerable pasture improvement the permanent grass, in conjunction with 60 hectares of 2-year leys, now carries 150 Aberdeen Angus × Friesian suckler cows—80 calving in the autumn and 70 calving in the winter. A herd of 30 pedigree Charolais cows breed good quality performance-tested bulls for the suckler and dairy markets. In addition, a ewe flock of 650 mules produces lambs which are finished on the farm.

The development of the suckler herds at Givendale has been built on strict management controls which were intended to maximise performance at every stage of production. Particular emphasis has been put on maintaining a tight 6-week calving period for each herd and on achieving good calf performance. The progress of the herds has been kept on target with the help of the MLC Beefplan recording service.

FIGURE 9.1 Plan of Givendale Farm showing the grazing, arable and wooded areas

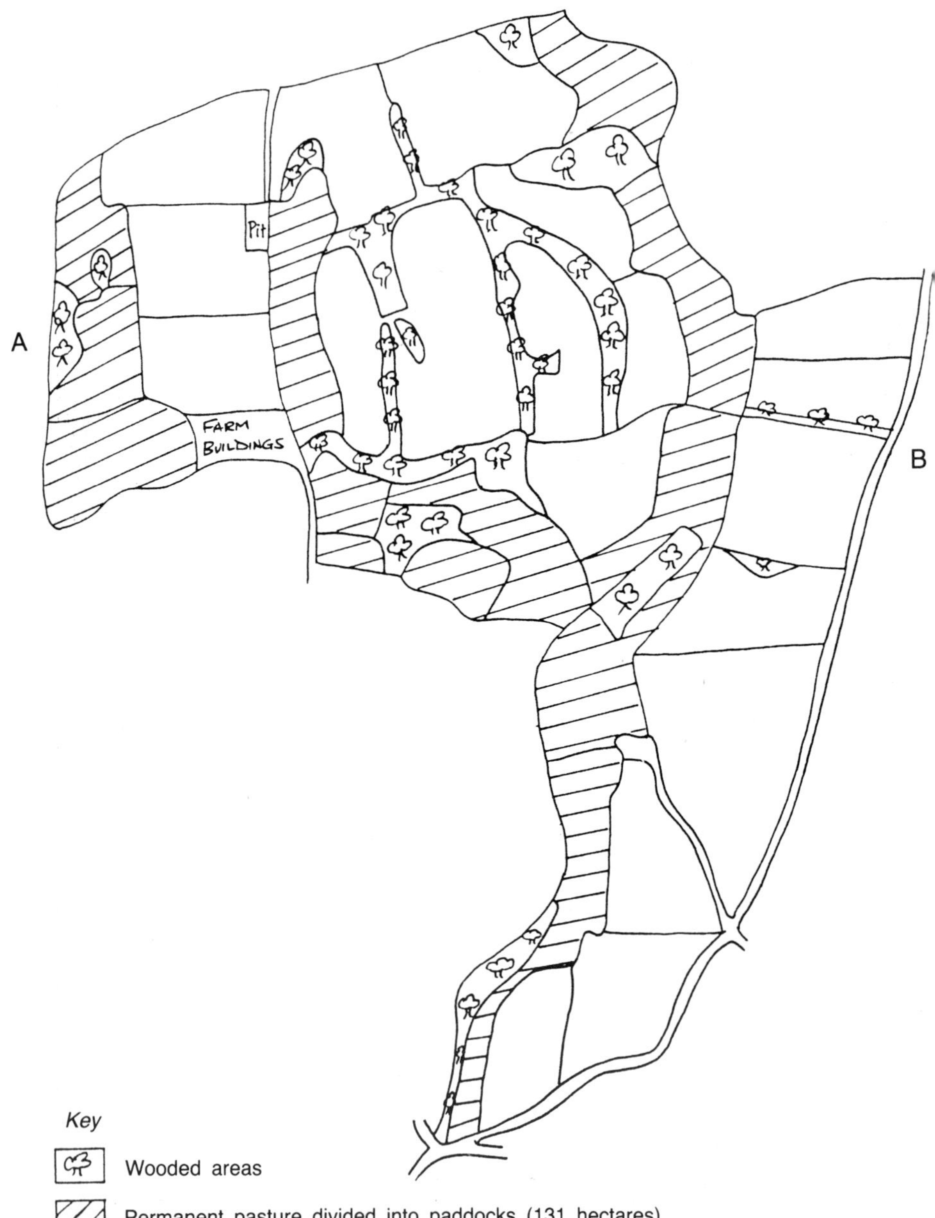

Key

 Wooded areas

Permanent pasture divided into paddocks (131 hectares)

Arable land (271 hectares) and roads

For profile from A to B (1.5 miles) see Figure 9.2.

FIGURE 9.2 Topography of Givendale Farm in relation to land use. Profile from A to B, (see Figure 9.1)

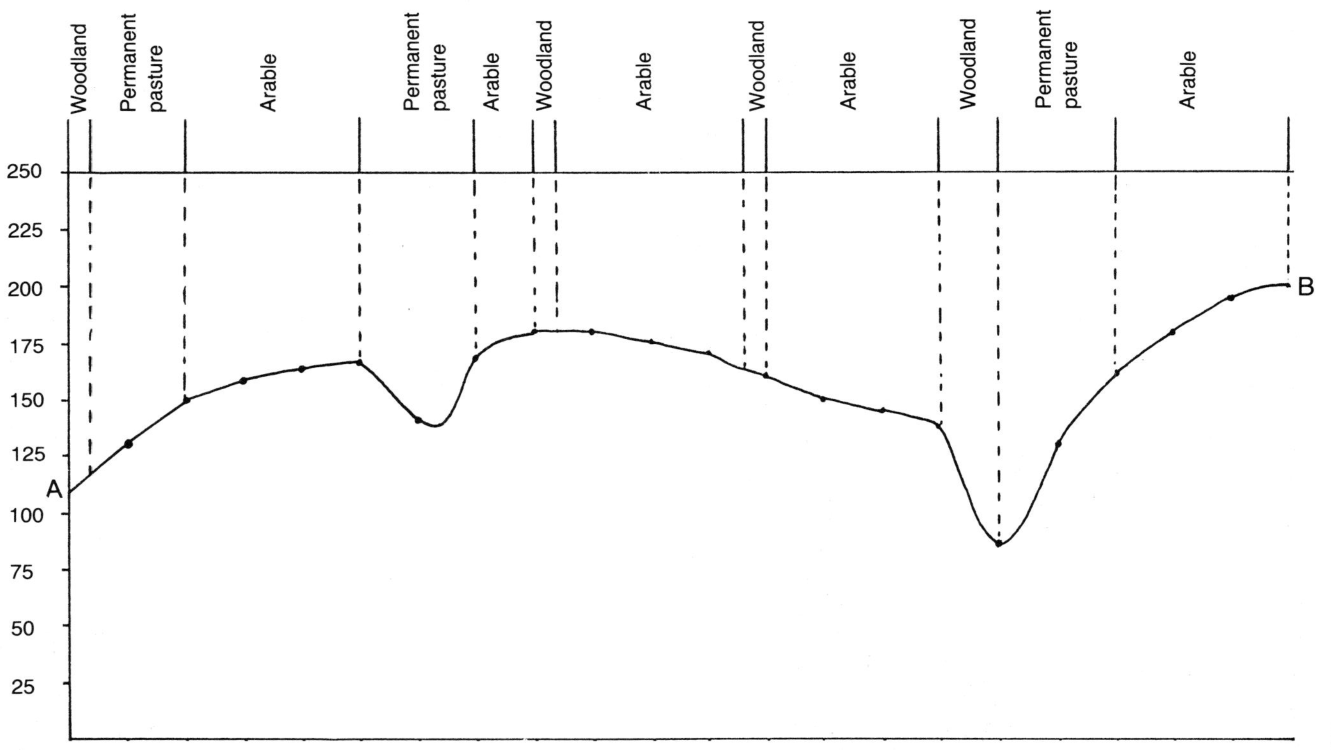

THE GRAZING SYSTEM

When the farm was acquired the grassland was heavily infested with rabbits, and the presence of large areas of scrub considerably reduced the area of productive grazing. In addition, much of the fencing was dilapidated and had to be replaced. The improvement plan for the farm emphasised the need to maximise the stock-carrying capacity of the grassland, with ease of management a priority.

The scrub and the rabbit burrows were bulldozed and the areas of grassland were re-fenced to create paddocks of about 12 hectares in size, as shown in Figure 9.1. Some reseeding of the areas which were damaged by the bulldozers was necessary but, in the main, pasture improvement has been achieved by encouraging the indigenous grasses with the use of fertilisers, sprays and close grazing. The division of the grassland into 12-hectare paddocks has enabled a controlled, clean grazing system to operate by alternating grazing between the cattle and sheep every other year. The paddock system has also proved essential in the establishment of bull beef production from the suckler herds.

Groups of cattle and sheep are set-stocked at about 1 cow and calf per acre or 5 ewes and lambs per acre. The intensity of stocking varies slightly, depending on the grass production potential of each area. Fertilisers have played an important role in improving pasture productivity. The first dressing of nitrogen fertiliser (90 kg/ha) is applied in early March by aeroplane. The second application of 90 kg per hectare is spread at the end of May by a 4 wheeled drive tractor. A further 40 kg per hectare is applied at the end of June, and a 15-15-15 compound is then applied in early August at the rate of 90 kg per hectare to maintain the P and K levels and to ensure autumn grass growth.

Weed control in the grassland is considered very important. Chemical sprays are used with care to control problem weeds such as nettles and thistles.

FORAGE

Temporary 2-year leys provide silage for the cows as well as grazing for the weaned lambs from the ewe flock. Leys are established by drilling into a properly ploughed and worked seed bed, after winter barley in early August. Tetraploid ryegrasses are grown and three cuts of silage are taken before the end of July, when lambs begin to graze the leys. A total of 375 kg per hectare of nitrogen (N) is applied during the season. The grass is cut with a trailed mower-conditioner and is then picked up after a 24-hour wilt with a multi-chop machine.

The silage clamp, which holds 1,000 tonnes, was constructed outside by digging into an earth bank and laying a concrete base. The target is to make material with a 'D' value of 67, dry matter (DM) of 25 percent and an energy content of 10.8 MJ ME/kg DM.

The other form of forage fed at Givendale is wheat straw treated with anhydrous ammonia. This has proved to be an excellent feed for dry cows, replacement heifers and pedigree Charolais cows. The availability of large quantities of wheat straw, coupled with economic treatment of 30-tonne stacks has made the product a useful alternative to silage.

SEASON OF CALVING

Due to the availability of 2-year leys it was decided that a high input, high output, silage-based suckler system would best fit the resources of the farm. Further considerations, such as the types of buildings that were in existence on the farm, contributed to the decision to calve 80 cows in the autumn and 70 cows in the winter. In addition, by splitting the calving season the same herd sires could be used twice in one year.

COW TYPE

Aberdeen Angus × Friesian cows were chosen because of their breeding efficiency, particularly in relation to holding to service, and because of their ability to produce fast growing, good quality calves. Although the future policy for the herd is to continue to use mainly Angus × Friesians, it is recognised that other breeds such as Simmental × and Limousin × Friesians should be considered. It is therefore likely that these breeds will be introduced on a trial basis.

CULLING AND HERD REPLACEMENTS

Cows are culled when they are 10 years old or when they develop physical problems such as infertility, or diseased or pendulous udders. The herd is replaced at a rate of 14 percent per year. Heifers, selected and bought as yearlings, are mated when they weigh at least 350 kg, using a Limousin bull. They calve down 6 weeks before the cows, allowing more time for post-calving recovery and giving more time before weaning for their smaller calves to grow. They are always housed separately to allow better control of feeding in order to bring them to the correct condition for successful re-breeding.

Autumn born bull calves with their dams in early May at Givendale.

HERD SIRES

All the cows are mated to Charolais bulls because, as discussed in an early chapter, calves sired by Charolais tend to grow faster and to reach heavier weights than calves sired by other breeds.

Bulls are selected from the Givendale pedigree Charolais herd for ease of calving, good fleshing ability, fast growth rate and sound conformation. The selection of bulls is based on group performance testing, and those with the best results are identified for use on the suckler cows. The bulls are then progeny tested to verify their true breeding value, and the information which is gathered is used to improve the pedigree breeding programme. Surplus young bulls are sold as herd sires to suckler and dairy units.

THE AUTUMN CALVING HERD

Cows calve outside from mid-September to the end of October, with 85 percent of the herd calving in the first 4 weeks. Shortly before calving is due to start all the cows are moved into an 8 hectare paddock and magnesium syrup is made available to prevent grass staggers. The herd is observed every 3 hours around the clock, with night patrols being carried out using a Land Rover which is equipped with a spot light. A careful watch is kept on calving cows, and assistance is given when necessary in a wooden catching pen which has been constructed in the calving field. A calving aid is always available.

The calves are tagged soon after they are born to correspond with their mothers' numbers. They are then moved into adjacent paddocks where they remain until they are housed. The herd stays out until the end of October and straw is usually fed to the cows so that dry matter intake is maintained. When the cows and calves are housed, the yarding system allows the grouping together of replacement heifers, the second calvers and thin cows, and the fitter mature cows. Yarding the herd in these 3 groups enables feeding to be regulated according to the nutritional requirements of the differing classes of cow.

The herd is fed silage ad libitum with increasing quantities of rolled barley up to the 6 week mating period, by which time the heifers and the younger cows are receiving 2.5 kg barley per day and the mature cows 1.8 kg per day. In addition, a high phosphorus mineral (10% P) is rationed by sprinkling it on to the silage. This feeding regime is intended to provide the cows with a rising plane of nutrition to ensure that good conception rates are achieved. The target is for about 96 percent of the cows to hold in calf during the 6 week mating period. To achieve this standard the cows must not only be well-fed but they must also be healthy, so they are wormed towards the end of November with a broad spectrum anthelmintic; they are also treated for lice.

In February the cows with the bull calves are split from the cows with heifer calves and they stay as separate groups during the grazing and finishing periods. During February the high phosphorus mineral is replaced by a general purpose one and the daily ration of barley is slowly reduced. The rate of reduction depends on the condition of the cows and on the quality of the silage.

The calves are creep-fed during the winter, up to a maximum of 2.5 kg per head per day. The creep feed is a mixture of rolled barley and sugar beet pulp, with a protein balancer added. The target weights for the calves by the time they are turned out to grass are 320 kg for the bulls and 290 kg for the heifers.

Before the herd is turned out during the last week in April, the cows are pregnancy tested. Any barren cows can either be re-mated to come into the winter calving herd or they can be earmarked for culling after they have put on condition at grass. Barren cows are never turned out with the cows which have bull calves at foot. When the herd is turned out to grass the cows and calves are set stocked on the grazing areas. Liquid magnesium syrup is made available to the cows in ball feeders for 6 weeks, to counteract the risk of hypomagnesaemia.

The calves grow well at grass, with the heifers gaining about 1 kg per day and the bulls about 1.2 kg a day. These gains are achieved on grass and milk only. When the calves are weaned at the end of July they are weighed, wormed and then housed for finishing.

The cows are treated with a dry-cow antibiotic, their teats are dipped and they are then housed for 3 days and fed on straw. Liquid magnesium syrup is also fed at this stage to help the cows to digest the straw and to counteract the possibility of hypomagnesaemia associated with stress. When the cows are turned out again they are moved onto bare pasture, at a high stocking rate. This prevents them from becoming too fat before they start to calve again in the middle of September.

The management of the autumn calving herd over the year is summarised in Figure 9.3.

THE WINTER CALVING HERD

The dry cows are brought inside when the weather deteriorates in November. They are housed in a cubicle shed and fed anhydrous ammonia-treated wheat straw supplemented with a special high sulphur mineral. The second calvers are usually penned separately and fed 1.5 kg of rolled barley per day to improve their body condition. The cows are wormed with a broad spectrum anthelmintic and they are also treated for lice. All the cows and heifers are vaccinated to prevent their calves from contracting rotavirus or E. coli infections. Vaccination takes place about a month before calving is due and the preventative programme has proved to be very successful.

On 1 February, shortly before the cows are due to calf, they are moved out of the cubicle shed into fold yards which were previously occupied by autumn calvers. The latter are moved into the cubicle sheds. As the cows start to calve, their nutritional requirements increase so silage replaces the treated wheat straw. Calving is monitored very carefully and the herd is seen every 3 hours round the clock. The cows are penned in groups of 25 as they calve, and the yards are kept well bedded with clean straw.

FIGURE 9.3 The autumn calving herd calendar

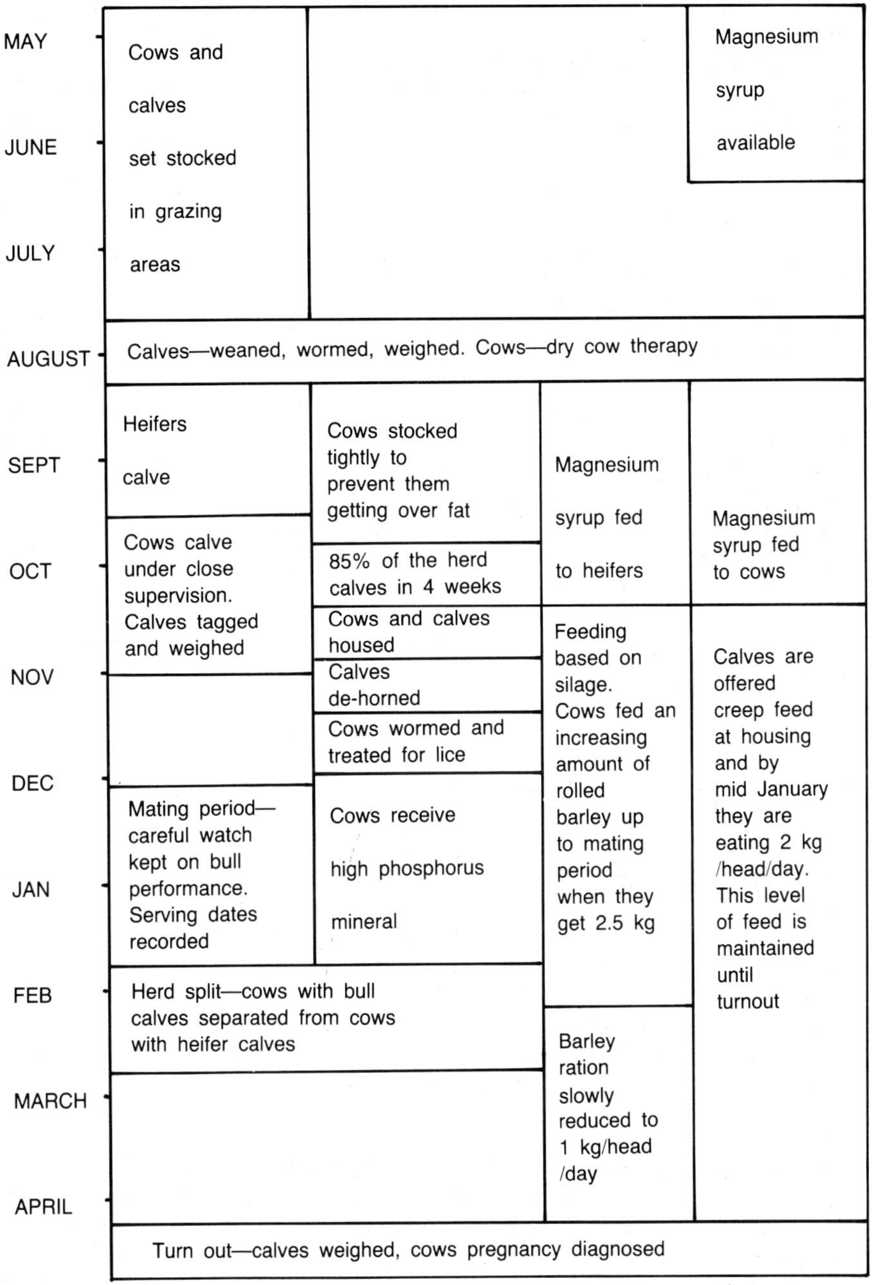

MAY	Cows and calves set stocked in grazing areas			Magnesium syrup available
JUNE				
JULY				
AUGUST	Calves—weaned, wormed, weighed. Cows—dry cow therapy			
SEPT	Heifers calve	Cows stocked tightly to prevent them getting over fat	Magnesium syrup fed to heifers	Magnesium syrup fed to cows
OCT	Cows calve under close supervision. Calves tagged and weighed	85% of the herd calves in 4 weeks		
NOV		Cows and calves housed	Feeding based on silage. Cows fed an increasing amount of rolled barley up to mating period when they get 2.5 kg	Calves are offered creep feed at housing and by mid January they are eating 2 kg /head/day. This level of feed is maintained until turnout
		Calves de-horned		
		Cows wormed and treated for lice		
DEC	Mating period—careful watch kept on bull performance. Serving dates recorded	Cows receive high phosphorus mineral		
JAN				
FEB	Herd split—cows with bull calves separated from cows with heifer calves		Barley ration slowly reduced to 1 kg/head /day	
MARCH				
APRIL				
	Turn out—calves weighed, cows pregnancy diagnosed			

As soon as the calves are born their navels are soaked in strong iodine solution and the treatment is repeated the next day. Iodine has proved to be a good material for sterilising and drying the umbilical cord quickly, preventing navel infections. It is, of course, essential for every calf to get the right amount of colostrum within 6 hours of birth and all calves are checked to make sure that they suckle their mothers properly. The calves are watched very carefully for any signs of scouring or pneumonia so that treatment can be implemented swiftly if necessary. We aim to rear healthy calves that will grow to their full potential unhindered by disease.

The cows are fed on good quality silage, which is sufficient to provide for maintenance and lactation until they are turned out to grass at the end of April. Heifers, however, are fed 2 kg per day of rolled barley in addition to silage. Before turn-out the calves are dis-budded and both the cows and the calves are given copper needles to supplement copper-deficient pastures. A high phosphorus mineral is given to the cows 3 weeks prior to the start of the mating period, which coincides with the herd being turned out to grass. The herd is split into two groups and a bull is run with each herd for three weeks, after which time the herds are joined together and a third bull is turned in for a further three weeks. To counteract the risk of hypomagnesaemia, magnesium syrup is made available to the cows for the first 6 weeks of the grazing season.

In early July the cows with the bull calves are separated from the cows with the heifer calves, and the two groups are then grazed in different paddocks. At this stage creep feed is made available to the calves, in hopper feeders, allowing one 5-space feeder for 15 to 20 calves. Initially the calves take small quantities of creep but by the end of August it needs to be restricted to 2.5 kg per calf per day. If grass runs short in the autumn the cows are buffer-fed with either spring barley straw or ammonia-treated straw. Magnesium syrup is fed to the cows from about 1 September until the calves are weaned in the middle of October.

At weaning, the calves are weighed and wormed, then housed for finishing. Dry cow therapy is practised and the cows are pregnancy tested. Those found not to be in calf are sold if they are fit enough, or they are housed for a fattening period. After a housed, 3-day drying off period the in-calf cows are turned out again for about another month before they are housed for the winter.

Figure 9.4 summarises the annual calendar for the winter calving herd.

FIGURE 9.4 The winter calving herd calendar

Month			
MAY – JUNE	Cows and calves set stocked in grazing areas	Mating period – careful watch on bull performance	Magnesium syrup fed
JULY – AUG		Cows with bull calves are split from cows with heifer calves.	
SEPT – OCT		Creep feed fed to calves	Magnesium syrup fed
OCT	Calves—weaned, wormed, weighed and housed Cows—dry cow therapy and pregnancy diagnosed		
NOV	Cows outside		
DEC – JAN	Cows housed and fed ammonia treated wheat straw	Cows and heifers wormed Heifers vaccinated with Rotavac Cows vaccinated with Rotavac	
JAN – FEB		Heifers calve	Health status of calves watched carefully. Particular attention paid to — colostrum — navel dressing — scours — pneumonia. Copper needles given and disbudding carried out before turn-out
FEB – MARCH – APRIL	Cows fed silage and a high-phosphorus mineral. Copper needles are given before turn-out	Cows calve under close supervision. Calves tagged and weighed	

IMPROVEMENTS IN PERFORMANCE

Table 9.1 shows the average calf weaning weights which were achieved at Givendale during the 10-year period from 1978 to 1987. The improved performance of the calves was the result of putting into practice the management techniques which have been outlined in this book.

TABLE 9.1 Average calf weaning weights at Givendale from 1978 to 1987

	Calf weaning weight (kg)									
	1978	*1979*	*1980*	*1981*	*1982**	*1983*	*1984*	*1985*	*1986*	*1987*
Autumn herd (Weaned at 10 months)										
Charolais × bulls	312	300	322	341	361	406	390	404	418	424
Charolais × heifers	288	278	305	315	340	349	357	364	360	365
Winter herd (Weaned at 7 months)										
Charolais × bulls	281	294	310	322	310	317	309	350	348	351
Charolais × heifers	233	260	275	296	274	277	270	284	298	310

* 1982 was the first year that male calves were left entire

When the gross margins per cow for the herd recorded by the MLC are considered (Table 9.2) the increases which were made from 1981 to 1987 only just kept pace with inflation. This highlights the static state the industry has been in over the last few years and demonstrates the need to increase output just to maintain margins.

TABLE 9.2 Gross margins per cow recorded by the MLC at Givendale from 1981 to 1987

	Gross margin (£/cow)						
	1981	*1982*	*1983*	*1984*	*1985*	*1986*	*1987*
Autumn herd	226	217	284	261	257	275	258
Spring herd	191	201	205	219	214	245	246

THE FUTURE

It is not easy to predict accurately future trends in beef production and consumption, but the evidence now available suggests that the supply of good quality beef is likely to decline. A culmination of factors seem to have brought about the reversal of the past trend of over production and 'beef mountains'.

Firstly, European veal consumption is increasing and part of this demand is being met by British calves. Therefore large numbers of potential beef cattle are being taken out of the beef production chain. Secondly, dairy cow numbers are being reduced all over Europe in response to milk quotas. This reduction has a profound effect because 60 percent of beef comes from the dairy herd. Thirdly, the poor conformation of Holstein-bred beef cattle frustrates the meat trade, which is increasingly discriminating against this type of beef by offering low prices.

With respect to consumption, beef is the most popular red meat and as national prosperity increases so beef is likely to be in greater demand. It therefore appears at present that the future of good quality suckled beef is assured.

Many farmers, quite correctly, are aware of this and, faced with the prospect of cereal set-aside, they are starting suckler herds as alternative enterprises. The growing enthusiasm for suckler systems should be tempered with words of caution.

I must stress that profitable suckled calf production is not the easy- going pursuit that many farmers might think it is. The degree of profitability depends on breeders being able to organise the management of their herds with the professional commitment which is necessary to make the business successful.

The overriding consideration, however, when contemplating a suckler herd is the availability of *capital*. At present, established herds performing to the highest standards are finding it difficult to produce a return which is higher than the current bank rate. It is therefore of paramount importance to budget carefully, with professional help, to try to ensure an acceptable return on the capital involved.

New entrants with limited experience of managing a breeding herd will have an uphill struggle as they compete for expensive breeding stock, with no real guarantee of a profitable beef price at the end of the day. On the other hand there are many existing breeders who will have to tighten their management standards if they are to have a certain future in beef production.

At Givendale further expansion of cattle numbers will be restricted to the pedigree Charolais herd. To meet the increasing demand for good quality Givendale performance tested Charolais bulls, the present herd of 35 cows will be increased to 50, and an annual embryo transfer program will increase the output of progeny. As far as the commercial suckler herds are concerned their future profitability will depend on better beef prices and on continuing to refine management techniques.